Measuring and Data

From Paces to Feet

Grade 3

Also appropriate for Grade 4

Karen Economopoulos
Jan Mokros
Rebecca B. Corwin
Susan Jo Russell

Developed at TERC, Cambridge, Massachusetts

Dale Seymour Publications®

The *Investigations* curriculum was developed at TERC (formerly Technical Education Research Centers) in collaboration with Kent State University and the State University of New York at Buffalo. The work was supported in part by National Science Foundation Grant No. MDR-9050210. TERC is a nonprofit company working to improve mathematics and science education. TERC is located at 2067 Massachusetts Avenue, Cambridge, MA 02140.

This project was supported, in part, by the
National Science Foundation
Opinions expressed are those of the authors and not necessarily those of the Foundation

This book is published by Dale Seymour Publications®, an imprint of the Alternative Publishing Group of Addison-Wesley Publishing Company.

Project Editor: Priscilla Cox Samii
Series Editor: Beverly Cory
ESL Consultant: Nancy Sokol Green
Production/Manufacturing Director: Janet Yearian
Production/Manufacturing Coordinator: Barbara Atmore
Design Manager: Jeff Kelly
Design: Don Taka
Illustrations: DJ Simison, Carl Yoshihara
Cover: Bay Graphics
Composition: Publishing Support Services

Printed on Recycled Paper

This book is based on *Measuring: From Paces to Feet* by Rebecca B. Corwin and Susan Jo Russell, which was originally published by Dale Seymour Publications as part of the series *Used Numbers: Real Data in the Classroom,* copyright © 1990 by Dale Seymour Publications®.

Order number DS21241
ISBN 0-86651-803-7
5 6 7 8 9 10-ML-00 99 98 97 96

INVESTIGATIONS IN NUMBER, DATA, AND SPACE

Principal Investigator **Susan Jo Russell**
Co-Principal Investigator **Cornelia C. Tierney**
Director of Research and Evaluation **Jan Mokros**

Curriculum Development
Joan Akers
Michael T. Battista
Mary Berle-Carman
Douglas H. Clements
Karen Economopoulos
Ricardo Nemirovsky
Andee Rubin
Susan Jo Russell
Cornelia C. Tierney
Amy Shulman Weinberg

Evaluation and Assessment
Mary Berle-Carman
Abouali Farmanfarmaian
Jan Mokros
Mark Ogonowski
Amy Shulman Weinberg
Tracey Wright
Lisa Yaffee

Teacher Development and Support
Rebecca B. Corwin
Karen Economopoulos
Tracey Wright
Lisa Yaffee

Technology Development
Michael T. Battista
Douglas H. Clements
Julie Sarama Meredith
Andee Rubin

Video Production
David A. Smith

Administration and Production
Amy Catlin
Amy Taber

Graduate Assistants
Kent State University:
Joanne Caniglia, Pam DeLong, Carol King
State University of New York at Buffalo:
Rosa Gonzalez, Sue McMillen,
Julie Sarama Meredith, Sudha Swaminathan

Cooperating Classrooms for This Unit
Katie Bloomfield
Robert A. Dihlmann
Shutesbury Elementary, Shutesbury, MA

Connie Brady
Laurie Friedlander
New York City Public Schools, New York, NY

Dolly Davis
Jane Holman
Clarke County Public Schools, Georgia

Eric Johnson
Boston Public Schools, Boston, MA

Jeanne Wall
Arlington Public Schools, Arlington, MA

Susan Wheelwright
Lucy Wittenberg
Fayerweather Street School, Cambridge, MA

Consultants and Advisors
Elizabeth Badger
Deborah Lowenberg Ball
Marilyn Burns
Ann Grady
Joanne M. Gurry
James J. Kaput
Steven Leinwand
Mary M. Lindquist
David S. Moore
John Olive
Leslie P. Steffe
Peter Sullivan
Grayson Wheatley
Virginia Woolley
Anne Zarinnia

CONTENTS

Teacher Notes

ABOUT THE *INVESTIGATIONS* CURRICULUM

Investigations in Number, Data, and Space is a K–5 mathematics curriculum with four major goals:

- to offer students meaningful mathematical problems
- to emphasize depth in mathematical thinking rather than superficial exposure to a series of fragmented topics
- to communicate mathematics content and pedagogy to teachers
- to substantially expand the pool of mathematically literate students

The *Investigations* curriculum embodies an approach radically different from the traditional textbook-based curriculum. At each grade level, it consists of a set of separate units, each offering 2–4 weeks of work. These units of study are presented through investigations that involve students in the exploration of major mathematical ideas.

Approaching the mathematics content through investigations helps students develop flexibility and confidence in approaching problems, fluency in using mathematical skills and tools to solve problems, and proficiency in evaluating their solutions. Students also build a repertoire of ways to communicate about their mathematical thinking, while their enjoyment and appreciation of mathematics grows.

The investigations are carefully designed to invite all students into mathematics—girls and boys, diverse cultural, ethnic, and language groups, and students with different strengths and interests. Problem contexts often call on students to share experiences from their family, culture, or community. The curriculum eliminates barriers—such as work in isolation from peers, or emphasis on speed and memorization—that exclude some students from participating successfully in mathematics. The following aspects of the curriculum ensure that all students are included in significant mathematics learning:

- Students spend time exploring problems in depth.
- They find more than one solution to many of the problems they work on.
- They invent their own strategies and approaches, rather than relying on memorized procedures.
- They choose from a variety of concrete materials and appropriate technology, including calculators, as a natural part of their everyday mathematical work.
- They express their mathematical thinking through drawing, writing, and talking.
- They work in a variety of groupings—as a whole class, individually, in pairs, and in small groups.
- They move around the classroom as they explore the mathematics in their environment and talk with their peers.

While reading and other language activities are typically given a great deal of time and emphasis in elementary classrooms, mathematics often does not get the time it needs. If students are to experience mathematics in depth, they must have enough time to become engaged in real mathematical problems. We believe that a minimum of five hours of mathematics classroom time a week—about an hour a day—is critical at the elementary level. The plan and pacing of the *Investigations* curriculum is based on that belief.

For further information about the pedagogy and principles that underlie these investigations, see Teacher Notes throughout the units and the following books:

- *Implementing the* Investigations in Number, Data, and Space™ *Curriculum*
- *Beyond Arithmetic*

The *Investigations* curriculum is presented through a series of teacher books, one for each unit of study. These books not only provide a complete mathematics curriculum for your students, they offer materials to support your own professional development. You, the teacher, are the person who will make this curriculum come alive in the classroom; the book for each unit is your main support system.

While reproducible resources for students are provided, the curriculum does not include student books. Students work actively with objects and experiences in their own environment and with a variety of manipulative materials and technology, rather than with workbooks and worksheets filled with problems. We also make extensive use of the overhead projector as a way to present problems, to focus group discussion, and to help students share ideas and strategies. If an overhead projector is available, we urge you to try it as suggested in the investigations.

Ultimately, every teacher will use these investigations in ways that make sense for his or her particular style, the particular group of students, and the constraints and supports of a particular school environment. We have tried to provide with each unit the best information and guidance for a wide variety of situations, drawn from our collaborations with many teachers and students over many years. Our goal in this book is to help you, as a professional educator, implement this mathematics curriculum in a way that will give all your students access to mathematical power.

Investigation Format

The opening two pages of each investigation help you get ready for the student work that follows. Here you will read:

What Happens—a synopsis of each session or block of sessions.

Mathematical Emphasis—the most important ideas and processes students will encounter in this investigation.

What to Plan Ahead of Time— materials to gather, student sheets to duplicate, transparencies to make, and anything else you need to do before starting.

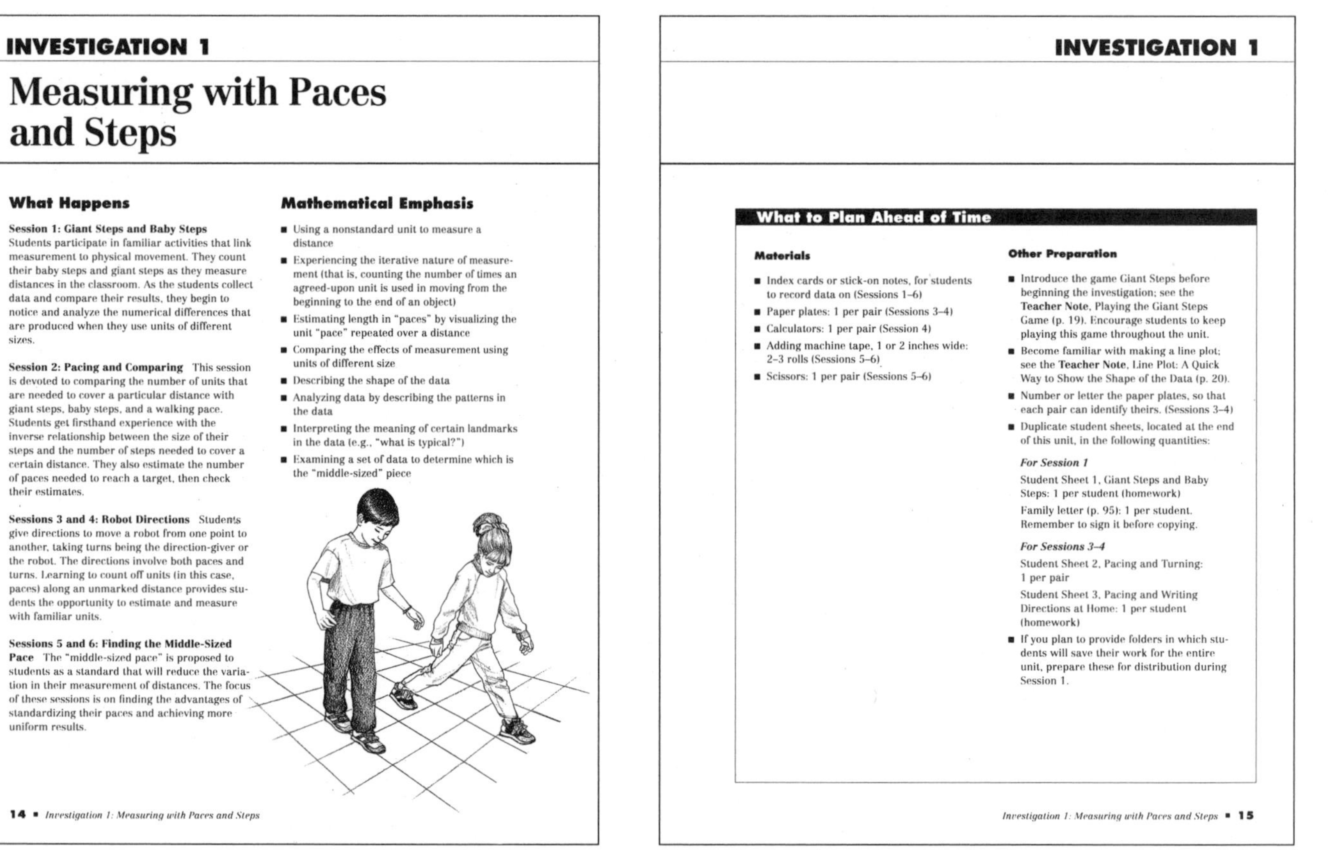
INVESTIGATION 1

Measuring with Paces and Steps

What Happens

Session 1: Giant Steps and Baby Steps Students participate in familiar activities that link measurement to physical movement. They count their baby steps and giant steps as they measure distances in the classroom. As the students collect data and compare their results, they begin to notice and analyze the numerical differences that are produced when they use units of different sizes.

Session 2: Pacing and Comparing This session is devoted to comparing the number of units that are needed to cover a particular distance with giant steps, baby steps, and a walking pace. Students get firsthand experience with the inverse relationship between the size of their steps and the number of steps needed to cover a certain distance. They also estimate the number of paces needed to reach a target, then check their estimates.

Sessions 3 and 4: Robot Directions Students give directions to move a robot from one point to another, taking turns being the direction-giver or the robot. The directions involve both paces and turns. Learning to count off units (in this case, paces) along an unmarked distance provides students the opportunity to estimate and measure with familiar units.

Sessions 5 and 6: Finding the Middle-Sized Pace The "middle-sized pace" is proposed to students as a standard that will reduce the variation in their measurement of distances. The focus of these sessions is on finding the advantages of standardizing their paces and achieving more uniform results.

Mathematical Emphasis

- Using a nonstandard unit to measure a distance
- Experiencing the iterative nature of measurement (that is, counting the number of times an agreed-upon unit is used in moving from the beginning to the end of an object)
- Estimating length in "paces" by visualizing the unit "pace" repeated over a distance
- Comparing the effects of measurement using units of different size
- Describing the shape of the data
- Analyzing data by describing the patterns in the data
- Interpreting the meaning of certain landmarks in the data (e.g., "what is typical?")
- Examining a set of data to determine which is the "middle-sized" piece

14 ▪ *Investigation 1: Measuring with Paces and Steps*

INVESTIGATION 1

What to Plan Ahead of Time

Materials

- Index cards or stick-on notes, for students to record data on (Sessions 1–6)
- Paper plates: 1 per pair (Sessions 3–4)
- Calculators: 1 per pair (Session 4)
- Adding machine tape, 1 or 2 inches wide: 2–3 rolls (Sessions 5–6)
- Scissors: 1 per pair (Sessions 5–6)

Other Preparation

- Introduce the game Giant Steps before beginning the investigation; see the **Teacher Note**, Playing the Giant Steps Game (p. 19). Encourage students to keep playing this game throughout the unit.
- Become familiar with making a line plot; see the **Teacher Note**, Line Plot: A Quick Way to Show the Shape of the Data (p. 20).
- Number or letter the paper plates, so that each pair can identify theirs. (Sessions 3–4)
- Duplicate student sheets, located at the end of this unit, in the following quantities:

  For Session 1

  Student Sheet 1, Giant Steps and Baby Steps: 1 per student (homework)

  Family letter (p. 95): 1 per student. Remember to sign it before copying.

  For Sessions 3–4

  Student Sheet 2, Pacing and Turning: 1 per pair

  Student Sheet 3, Pacing and Writing Directions at Home: 1 per student (homework)
- If you plan to provide folders in which students will save their work for the entire unit, prepare these for distribution during Session 1.

Investigation 1: Measuring with Paces and Steps ▪ 15

Sessions Within an investigation, the activities are organized by class session, a session being a one-hour math class. Sessions are numbered consecutively through an investigation. Often several sessions are grouped together, presenting a block of activities with a single major focus.

When you find a block of sessions presented together—for example, Sessions 1, 2, and 3—read through the entire block first to understand the overall flow and sequence of the activities. Make some preliminary decisions about how you will divide the activities into three sessions for your class, based on what you know about your students. You may need to modify your initial plans as you progress through the activities, and you may want to make notes in the margins of the pages as reminders for the next time you use the unit.

Be sure to read the Session Follow-Up section at the end of the session block to see what homework assignments and extensions are suggested as you make your initial plans.

While you may be used to a curriculum that tells you exactly what each class session should cover, we have found that the teacher is in a better position to make these decisions. Each unit is flexible and may be handled somewhat differently by every teacher. While we provide guidance for how many sessions a particular group of activities is likely to need, we want you to be active in determining an appropriate pace and the best transition points for your class.

Ten-Minute Math At the beginning of some sessions, you will find Ten-Minute Math activities. These are designed to be used in tandem with the investigations, but not during the math hour. Rather, we hope you will do them whenever you have a spare 10 minutes—maybe before lunch or recess, or at the end of the day.

Ten-Minute Math offers practice in key concepts, but not always those being covered in the unit. For example, in a unit on using data, Ten-Minute Math might revisit geometric activities done earlier in the year. Complete directions for the suggested activities are included at the end of each unit. A compilation of Ten-Minute Math activities is also available as a separate book.

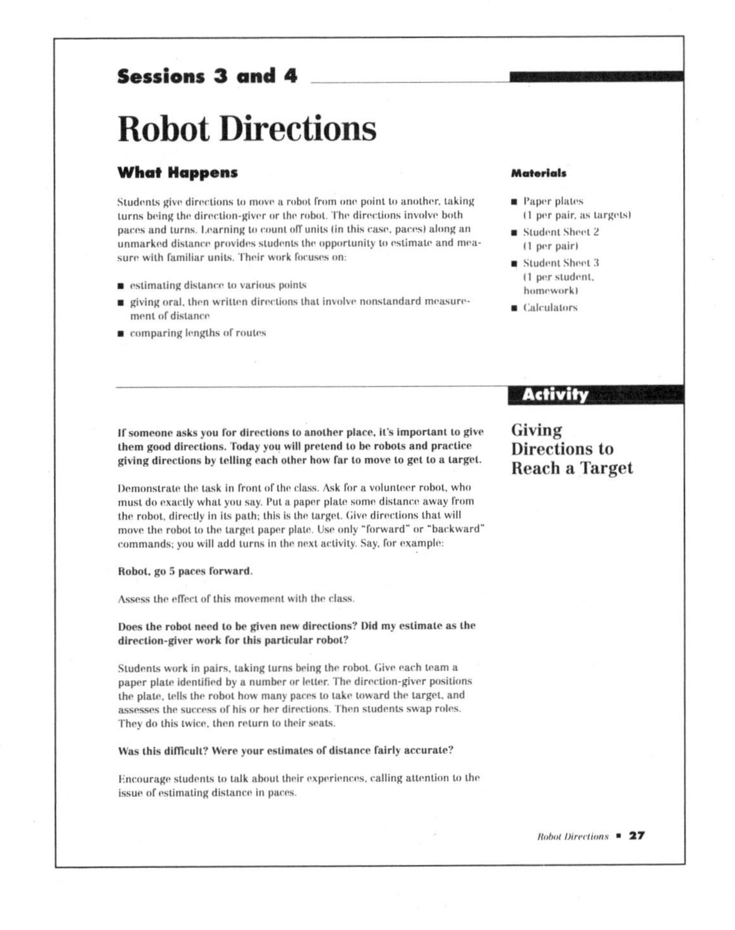

Sessions 3 and 4

Robot Directions

What Happens

Students give directions to move a robot from one point to another, taking turns being the direction-giver or the robot. The directions involve both paces and turns. Learning to count off units (in this case, paces) along an unmarked distance provides students the opportunity to estimate and measure with familiar units. Their work focuses on:

- estimating distance to various points
- giving oral, then written directions that involve nonstandard measurement of distance
- comparing lengths of routes

Materials

- Paper plates (1 per pair, as targets)
- Student Sheet 2 (1 per pair)
- Student Sheet 3 (1 per student, homework)
- Calculators

Activity

Giving Directions to Reach a Target

If someone asks you for directions to another place, it's important to give them good directions. Today you will pretend to be robots and practice giving directions by telling each other how far to move to get to a target.

Demonstrate the task in front of the class. Ask for a volunteer robot, who must do exactly what you say. Put a paper plate some distance away from the robot, directly in its path; this is the target. Give directions that will move the robot to the target paper plate. Use only "forward" or "backward" commands; you will add turns in the next activity. Say, for example:

Robot, go 5 paces forward.

Assess the effect of this movement with the class.

Does the robot need to be given new directions? Did my estimate as the direction-giver work for this particular robot?

Students work in pairs, taking turns being the robot. Give each team a paper plate identified by a number or letter. The direction-giver positions the plate, tells the robot how many paces to take toward the target, and assesses the success of his or her directions. Then students swap roles. They do this twice, then return to their seats.

Was this difficult? Were your estimates of distance fairly accurate?

Encourage students to talk about their experiences, calling attention to the issue of estimating distance in paces.

Robot Directions ■ **27**

Activities The activities include pair and small-group work, individual tasks, and whole-class discussions. We assume that students are seated together, talking and sharing ideas during all work times. Students most often work cooperatively, although each student may record work individually.

Choice Time In some units, some sessions are structured with activity choices. In these cases, students may work simultaneously on different activities focused on the same mathematical ideas. Students choose which activities they want to do, and they cycle through them.

You will need to decide how to set up and introduce these activities and how to let students make their choices. Some teachers present them as station activities, in different parts of the room. Some list the choices on the board as reminders or have students keep their own lists.

Tips for the Linguistically Diverse Classroom At strategic points in each unit, you will find concrete suggestions for simple modifications of the teaching strategies to encourage the participation of all students. Many of these tips offer alternative ways to elicit critical thinking from students at varying levels of English proficiency, as well as from other students who find it difficult to verbalize their thinking.

The tips are supported by suggestions for specific vocabulary work to help ensure that all students can participate fully in the investigations. The Preview for the Linguistically Diverse Classroom (p. 12) lists important words that are assumed as part of the working vocabulary of the unit. Second-language learners will need to become familiar with these words in order to understand the problems and activities they will be doing. These terms can be incorporated into students' second-language work before or during the unit. Activities that can be used to present the words and make them comprehensible are found in the appendix, Vocabulary Support for Second-Language Learners (p. 91).

In addition, ideas for making connections to students' language and cultures, included on the Preview page, help the class explore the unit's concepts from a multicultural perspective.

Session Follow-Up

Homework Homework is not given daily for its own sake, but periodically as it makes sense to have follow-up work at home. Homework may be used for (1) review and practice of work done in class; (2) preparation for activities coming up in class—for example, collecting data for a class project; or (3) involving and informing family members.

Some units in the *Investigations* curriculum have more homework than others, simply because it makes sense for the mathematics that's going on. Other units rely on manipulatives that most students won't have at home, making homework difficult. In any case, homework should always be directly connected to the investigations in the unit, or to work in previous units—never sheets of problems just to keep students busy.

Extensions These follow-up activities are opportunities for some or all students to explore a topic in greater depth or in a different context. They are not designed for "fast" students; mathematics is a multifaceted discipline, and different students will want to go further in different investigations. Look for and encourage the sparks of interest and enthusiasm you see in your students, and use the extensions to help them pursue these interests.

Family Letter A letter that you can send home to students' families is included with the blackline masters for each unit. We want families to be informed about the mathematics work in your classroom; they should be encouraged to participate in and support their children's work. A reminder to send home the letter appears in one of the early investigations. (These letters are also available separately in the following languages: Spanish, Vietnamese, Cantonese, Hmong, and Cambodian.)

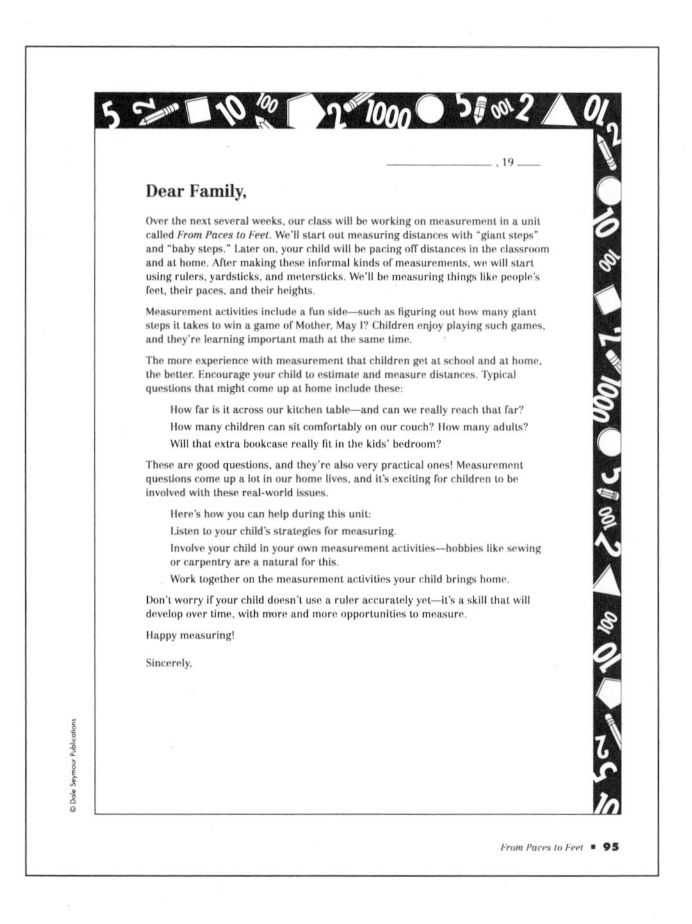

_______________, 19___

Dear Family,

Over the next several weeks, our class will be working on measurement in a unit called *From Paces to Feet*. We'll start out measuring distances with "giant steps" and "baby steps." Later on, your child will be pacing off distances in the classroom and at home. After making these informal kinds of measurements, we will start using rulers, yardsticks, and metersticks. We'll be measuring things like people's feet, their paces, and their heights.

Measurement activities include a fun side—such as figuring out how many giant steps it takes to win a game of Mother, May I? Children enjoy playing such games, and they're learning important math at the same time.

The more experience with measurement that children get at school and at home, the better. Encourage your child to estimate and measure distances. Typical questions that might come up at home include these:

How far is it across our kitchen table—and can we really reach that far?

How many children can sit comfortably on our couch? How many adults?

Will that extra bookcase really fit in the kids' bedroom?

These are good questions, and they're also very practical ones! Measurement questions come up a lot in our home lives, and it's exciting for children to be involved with these real-world issues.

Here's how you can help during this unit:

Listen to your child's strategies for measuring.

Involve your child in your own measurement activities—hobbies like sewing or carpentry are a natural for this.

Work together on the measurement activities your child brings home.

Don't worry if your child doesn't use a ruler accurately yet—it's a skill that will develop over time, with more and more opportunities to measure.

Happy measuring!

Sincerely,

Materials

A complete list of the materials needed for the unit is found on p. 10. Some of these materials are available in a kit for the *Investigations* grade 3 curriculum. Individual items can also be purchased as needed from school supply stores and dealers.

In an active mathematics classroom, certain basic materials should be available at all times: interlocking cubes, pencils, unlined paper, graph paper, calculators, things to count with, and measuring tools. Some activities in this curriculum require scissors and glue sticks or tape. Stick-on notes and large paper are also useful materials throughout.

So that students can independently get what they need at any time, they should know where these materials are kept, how they are stored, and how they are to be returned to the storage area. For example, interlocking cubes are best stored in towers of ten; then, whatever the activity, they should be returned to storage in groups of ten at the end of the hour. You'll find that establishing such routines at the beginning of the year is well worth the time and effort.

Student Sheets and Teaching Resources

Reproducible pages to help you teach the unit are found at the end of this book. These include masters for making overhead transparencies and other teaching tools, as well as student recording sheets.

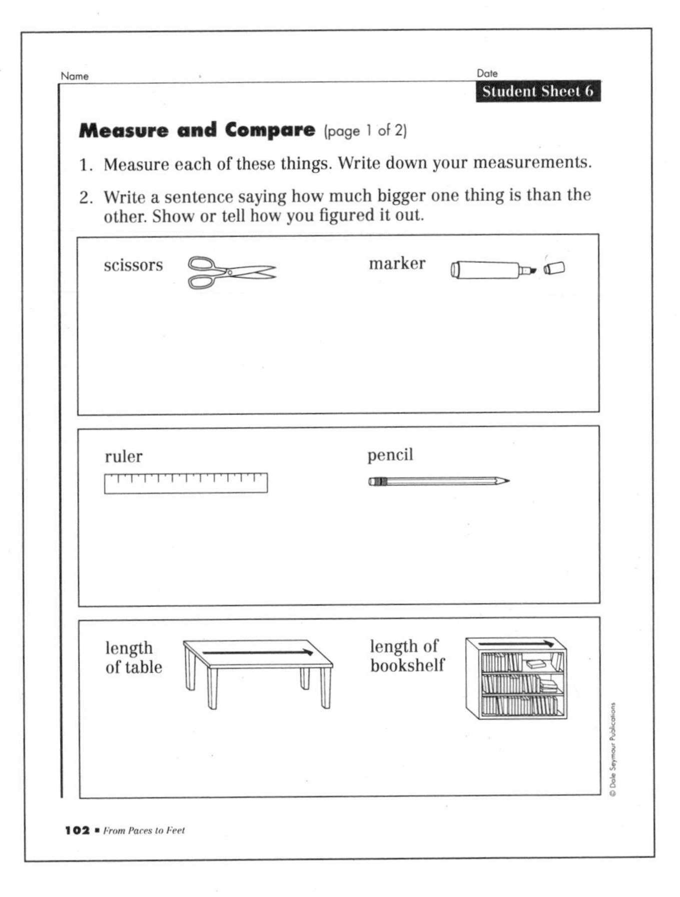

Name Date

Student Sheet 6

Measure and Compare (page 1 of 2)

1. Measure each of these things. Write down your measurements.
2. Write a sentence saying how much bigger one thing is than the other. Show or tell how you figured it out.

scissors marker

ruler pencil

length of table length of bookshelf

© Dale Seymour Publications

102 ▪ *From Paces to Feet*

Many of the field-test teachers requested more sheets to help students record their work, and we have tried to be responsive to this need. At the same time, we think it's important that students find their own ways of organizing and recording their work. They need to learn how to explain their thinking with both drawings and written words, and how to organize their results so someone else can understand them.

To ensure that students get a chance to learn how to represent and organize their own work, we deliberately do not provide student sheets for every activity. We recommend that your students keep a mathematics notebook or folder so that their work, whether on reproducible sheets or their own paper, is always available to them for reference.

Help for You, the Teacher

Because we believe strongly that a new curriculum must help teachers think in new ways about mathematics and about their students' mathematical thinking processes, we have included a great deal of material to help you learn more about both.

About the Mathematics in This Unit This introductory section (p. 11) summarizes for you the critical information about the mathematics you will be teaching. This will be particularly valuable to teachers who are accustomed to a traditional textbook-based curriculum.

Teacher Notes These reference notes provide practical information about the mathematics you are teaching and about our experience with how students learn. Many of the notes were written in response to actual questions from teachers, or to discuss important things we saw happening in the field-test classrooms. Some teachers like to read them all before starting the unit, then review them as they come up in particular investigations.

Dialogue Boxes Sample dialogues throughout the unit demonstrate how students typically express their mathematical ideas, what issues and confusions arise in their thinking, and how some teachers have guided class discussions.

These dialogues are based on the extensive classroom testing of this curriculum; many are word-for-word transcriptions of recorded class discussions. They are not always easy reading; sometimes it may take some effort to unravel what the students are trying to say. But this is the value of these dialogues; they offer good clues to how your students may develop and express their approaches and strategies, helping you prepare for your own class discussions.

Where to Start You may not have time to read everything the first time you use this unit. As a first-time user, you will likely focus on understanding the activities and working them out with your students. Read completely through each investigation before starting to present it.

When you next teach this same unit, you can begin to read more of the background. Each time you present this unit, you will learn more about how your students understand the mathematical ideas. The first-time user of *From Paces to Feet* should read the following:

- About the Mathematics in This Unit (p. 11)
- Line Plot: A Quick Way to Show the Shape of the Data (p. 20)
- The Shape of the Data: Clumps, Bumps, and Holes (p. 25)
- Connecting Oral and Written Directions (p. 31)
- Using Measuring Tools (p. 54)

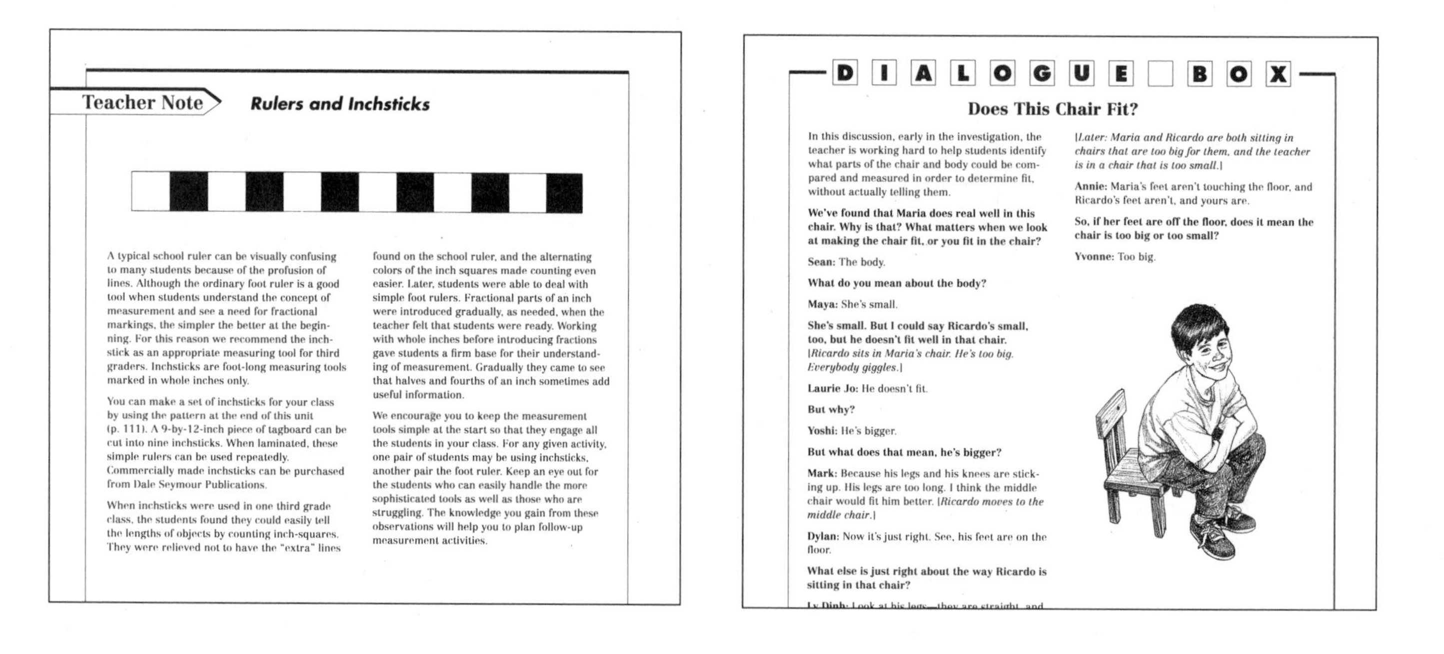

Teacher Note **Rulers and Inchsticks**

A typical school ruler can be visually confusing to many students because of the profusion of lines. Although the ordinary foot ruler is a good tool when students understand the concept of measurement and see a need for fractional markings, the simpler the better at the beginning. For this reason we recommend the inchstick as an appropriate measuring tool for third graders. Inchsticks are foot-long measuring tools marked in whole inches only.

You can make a set of inchsticks for your class by using the pattern at the end of this unit (p. 111). A 9-by-12-inch piece of tagboard can be cut into nine inchsticks. When laminated, these simple rulers can be used repeatedly. Commercially made inchsticks can be purchased from Dale Seymour Publications.

When inchsticks were used in one third grade class, the students found they could easily tell the lengths of objects by counting inch-squares. They were relieved not to have the "extra" lines found on the school ruler, and the alternating colors of the inch squares made counting even easier. Later, students were able to deal with simple foot rulers. Fractional parts of an inch were introduced gradually, as needed, when the teacher felt that students were ready. Working with whole inches before introducing fractions gave students a firm base for their understanding of measurement. Gradually they came to see that halves and fourths of an inch sometimes add useful information.

We encourage you to keep the measurement tools simple at the start so that they engage all the students in your class. For any given activity, one pair of students may be using inchsticks, another pair the foot ruler. Keep an eye out for the students who can easily handle the more sophisticated tools as well as those who are struggling. The knowledge you gain from these observations will help you to plan follow-up measurement activities.

DIALOGUE BOX

Does This Chair Fit?

In this discussion, early in the investigation, the teacher is working hard to help students identify what parts of the chair and body could be compared and measured in order to determine fit, without actually telling them.

We've found that Maria does real well in this chair. Why is that? What matters when we look at making the chair fit, or you fit in the chair?

Sean: The body.

What do you mean about the body?

Maya: She's small.

She's small. But I could say Ricardo's small, too, but he doesn't fit well in that chair. [*Ricardo sits in Maria's chair. He's too big. Everybody giggles.*]

Laurie Jo: He doesn't fit.

But why?

Yoshi: He's bigger.

But what does that mean, he's bigger?

Mark: Because his legs and his knees are sticking up. His legs are too long. I think the middle chair would fit him better. [*Ricardo moves to the middle chair.*]

Dylan: Now it's just right. See, his feet are on the floor.

What else is just right about the way Ricardo is sitting in that chair?

Ly Dinh: Look at his legs—they are straight, and

[*Later: Maria and Ricardo are both sitting in chairs that are too big for them, and the teacher is in a chair that is too small.*]

Annie: Maria's feet aren't touching the floor, and Ricardo's feet aren't, and yours are.

So, if her feet are off the floor, does it mean the chair is too big or too small?

Yvonne: Too big.

Teacher Checkpoints As a teacher of the *Investigations* curriculum, you observe students daily, listen to their discussions, look carefully at their work, and use this information to guide your teaching. We have designated Teacher Checkpoints as natural times to get an overall sense of how your class is doing in the unit.

The Teacher Checkpoints provide a time for you to pause and reflect on your teaching plan while observing students at work in an activity. These sections offer tips on what you should be looking for and how you might adjust your pacing. Are most students fluent with strategies for solving a particular kind of problem? Are they just starting to formulate good strategies? Or are they still struggling with how to start?

Depending on what you see as the students work, you may want to spend more time on similar problems, change some of the problems to use smaller numbers, move quickly to more challenging material, modify subsequent activities for some students, work on particular ideas with a small group, or pair students who have good strategies with those who are having more difficulty.

In *From Paces to Feet* you will find three Teacher Checkpoints:

Do Students' Directions Work? (p. 29)
Comparing Paces to the Standard (p. 35)
Observing as Students Measure (p. 59)

Embedded Assessment Activities Use the built-in assessments included in this unit to help you examine the work of individual students, figure out what it means, and provide feedback. From the students' point of view, the activities you will be using for assessment are no different from any others; they don't look or feel like traditional tests.

These activities sometimes involve writing and reflecting, at other times a brief interaction between student and teacher, and in still other instances the creation and explanation of a product.

In *From Paces to Feet* you will find three assessment activities in the second investigation to check on the development of students' skills in measuring and data description:

- The King's Foot (p. 43)
- Analyzing the Pattern Block Data (p. 63)
- Checking Students' Metric Sizes (p. 71)

Teachers find the hardest part of the assessment to be interpreting their students' work. If you have used a process approach to teaching writing, you will find our mathematics approach familiar. To help with interpretation, we provide guidelines and questions to ask about the students' work. In some cases we include a Teacher Note with specific examples of student work and a commentary on what it indicates. This framework can help you determine how your students are progressing.

As you evaluate students' work, it's important to remember that you're looking for much more than the "right answer." You'll want to know what their strategies are for solving the problem, how well these strategies work, whether they can keep track of and logically organize an approach to the problem, and how they make use of representations and tools to solve the problem.

Ongoing Assessment Good assessment of student work involves a combination of approaches. Some of the things you might do on an ongoing basis include the following:

- **Observation** Circulate around the room to observe students as they work. Watch for the development of their mathematical strategies, and listen to their discussions of mathematical ideas.
- **Portfolios** Ask students to document their work, in journals, notebooks, or portfolios. Periodically review this work to see how their mathematical thinking and writing are changing. Some teachers have students keep a notebook or folder for each unit, while others prefer one mathematics notebook or a portfolio of selected work for the entire year. Take time at the end of each unit for students to choose work for their portfolios. You might also have them write about what they've learned in the unit.

From Paces to Feet

OVERVIEW

Content of This Unit This unit explores both measurement and simple statistics, as students develop ideas about why we need to measure, learn to use different measuring tools and systems, and interpret data they collect by measuring. Through some initial work with informal, nonstandard units of measure (baby steps, giant steps, paces), students see that defining a standard or middle-sized pace provides more accurate and more consistent measures. Students then learn to use standard measuring tools—inch rulers and yardsticks, centimeter rulers and metersticks—as they collect measurement data about themselves and their classroom; they then learn ways to organize, represent, and analyze this data, discovering the power of measurement in communicating about the world.

Connections with Other Units This unit is intended for third graders, but could also be used with fourth graders who need more experiences with measurement. Students who have done a great deal of informal measurement (such as measuring with cubes or blocks) may spend less time on Investigation 1. However, take care not to underestimate the amount of time students need to spend with nonstandard units of measure. Many third graders are not yet ready to understand the rationale and uses of standard measurement. If this is the case in your class, spend more time on the first investigation.

Investigations Curriculum ■ Suggested Grade 3 Sequence

Mathematical Thinking at Grade 3 (Introduction)
Things That Come in Groups (Multiplication and Division)
Flips, Turns, and Area (2-D Geometry)
▶ *From Paces to Feet* (Measuring and Data)
Landmarks in the Hundreds (The Number System)
Up and Down the Number Line (Changes)
Combining and Comparing (Addition and Subtraction)
Turtle Paths (2-D Geometry)
Fair Shares (Fractions)
Exploring Solids and Boxes (3-D Geometry)

Investigation 1 • Measuring with Paces and Steps			
Class Sessions	**Activities**	**Pacing**	**Ten-Minute Math**
Session 1: GIANT STEPS AND BABY STEPS	Estimating Distance in Giant Steps Measuring in Baby Steps ■ Homework	1 hr	Estimation and Number Sense
Session 2: PACING AND COMPARING	Pacing the Classroom Estimating How Far in Paces ■ Extension	1 hr	
Sessions 3 and 4: ROBOT DIRECTIONS	Giving Directions to Reach a Target Pacing and Turning ■ Teacher Checkpoint: Do Students' Directions Work? Planning and Comparing Robot Routes ■ Homework ■ Extension	2 hrs	
Sessions 5 and 6: FINDING THE MIDDLE-SIZED PACE	What's the Middle-Sized Pace? ■ Teacher Checkpoint: Comparing Paces Using the Standard Pace ■ Extension	2 hrs	

Investigation 2 • From Paces to Feet			
Class Sessions	**Activities**	**Pacing**	**Ten-Minute Math**
Session 1: THE NEED FOR A STANDARD MEASURE	■ Assessment: The King's Foot Introduction to Measuring Tools ■ Homework ■ Extension	1 hr	Quick Images
Session 2: KIDS' FEET AND ADULTS' FEET	Looking at Foot Length Describing Adults' Foot Length Things Longer Than a Foot ■ Extension	1 hr	
Sessions 3 and 4: MEASURING CENTERS	Working at the Measuring Centers ■ Teacher Checkpoint: Observing as Students Measure Organizing Our Jumping Data Discussion: How Far Can a Third Grader Jump?	2 hrs	
Session 5: MOVING TO METRIC	■ Assessment: Analyzing the Pattern Block Data Introducing Centimeters and Meters A Scavenger Hunt ■ Homework	1 hr	
Sessions 6 and 7: METRIC MEASUREMENT	Discussing the Scavenger Hunt Body Size Measuring Centers Discussing Our Height Findings ■ Assessment: Checking Students' Metric Sizes	2 hrs	

Continued on next page

Investigation 3 • Measuring Project: Do Our Chairs Fit Us?			
Class Sessions	**Activities**	**Pacing**	**Ten-Minute Math**
Session 1: WHAT IS A GOOD FIT?	Looking at How Chairs Fit Measuring Ourselves and Our Chairs	1 hr	
Sessions 2 and 3: DO OUR CHAIRS FIT US?	Making a Clear Picture of the Data Making Our Report	2 hrs	

Investigation 4 • Measuring Project: Balobbyland			
Class Sessions	**Activities**	**Pacing**	**Ten-Minute Math**
Sessions 1, 2, and 3: MAKING A SMALL WORLD	Meeting the Balobbies Making Tree House Platforms Making More Balobby Spaces Planning a Neighborhood	3 hrs	

MATERIALS LIST

Following are the basic materials needed for the activities in this unit. The suggested quantities are ideal; however, in some instances you can work with smaller quantities by running several activities, requiring different materials, simultaneously.

Items marked with an asterisk are provided with the *Investigations* Materials Kit for grade 3.

* Rulers marked in inches and centimeters: 1 per pair of students

Inchsticks: 1 per pair of students

Yardsticks: 5 per class

Metersticks: 5 per class

Tape measure

How Big Is a Foot? by Rolf Myller (Dell, 1990) (optional)

* Pattern blocks: 5–6 trapezoids (checkers may be substituted)

* Centimeter cubes: 1 per pair of students (cutouts of one-centimeter squares, from graph paper, may be substituted) (optional)

* Adding machine tape, 1 or 2 inches wide: 2–3 rolls

Index cards or stick-on notes: 2–3 packs

Chart paper: 10 sheets

Paper plates: 1 per pair of students

Scissors: 1 per pair of students

Calculators: 1 per pair of students

Tape, glue, chalk, markers

Blank overhead transparencies and pens

Yarn or string

World map or globe

Materials needed only for Investigation 4

Toy people, 5–8 cm tall

Clay or playdough

Teddy bear counters (optional)

Tagboard

The following materials are provided at the end of this unit as blackline masters. They are also available in classroom sets.

Family Letter (p. 95)

Student Sheets 1–9 (pp. 97–106)

Teaching Resources:

Balobby Space Cards (pp. 107–110)

Inchstick Pattern (p. 111)

Centimeter Graph Paper (p. 112)

Quick Image Dot Patterns (p. 113)

Quick Image Geometric Designs (p. 114)

Linear measurement has long been a central topic in the elementary mathematics curriculum. Measurement connects mathematics to real life in powerful ways: It is a tool that we use to collect data and communicate about the world. This unit involves students in thinking about why people need to measure, the different tools and systems we use for measuring, and how we use and interpret data that are based on measurements.

Data collection and data analysis are important parts of this unit. Students collect real data through measuring, using both informal and standard measurement systems. They then represent the data in a variety of ways, describe landmarks and features of the data, and finally formulate hypotheses and build theories about the reality represented by the data. Through the activities in this unit, students learn to use simple statistical tools and concepts, gaining a good foundation for their later work in statistics and data analysis.

Researchers and educators, including Piaget, have extensively studied how children learn to measure. We know that young students' ideas about measurement grow out of a great deal of experience with informal measurement: Constructing how long something is in baby steps or in unit cubes is not just playing at measurement—it is an important mathematical construction. Seeing the need to describe length in a reliable and accurate way, so that you'll get the same results if you measure a second time or if someone else does it, is a skill that students develop through repeated experiences with describing and comparing the sizes of things.

Students need many opportunities to use informal as well as more standard measuring tools. As they use different units of measure, they begin to learn about the relationship between sizes of measuring units and the results of measuring (that is, for any given measure, the smaller the measuring unit, the greater the total amount of units needed). Through discussing the methods and units they use, they learn that defining procedures, recording information, and agreeing upon a standard are critical parts of communicating measurement information to others.

Mathematical Emphasis At the beginning of each investigation, the Mathematical Emphasis section tells you what is most important for students to learn about during that investigation. Many of these mathematical understandings and processes are difficult and complex. Students gradually learn more and more about each idea over many years of schooling. Individual students will begin and end the unit with different levels of knowledge and skill, but all will gain greater knowledge of using measuring tools and understanding what it means to measure, and will develop good beginning data analysis skills.

A Note on Measuring Tools The rulers available in different classrooms vary considerably. Those provided with the *Investigations* grade 3 materials kit are marked in inches on one side and centimeters on the other. Students will be using both systems of measurement at different times in this unit. If you find that your students are confusing the two sides, consider using masking tape to cover the side they are not working with.

For measures longer than 1 foot, some schools may have a combination tool—metersticks that are marked with inches on the reverse. These are fine for use with metric measure; however, be careful if you plan to use them for measuring in inches. They look like yardsticks, but they are actually a little more than 39 inches long. This can be confusing to students (and adults!) who use this tool, expecting to measure things in 3-foot lengths.

To reduce the confusion, try covering the extra 3 inches with masking tape when you need *yardsticks*. When you introduce these tools (Investigaton 2), explain why you have covered the end. If you can get separate yardsticks and metersticks, use these instead of the combination stick.

Foot rulers and 100-cm strips of paper tape can be used instead of yardsticks and metersticks throughout this unit.

In the *Investigations* curriculum, mathematical vocabulary is introduced naturally during the activities. We don't ask students to learn definitions of new terms; rather, they come to understand such words as *factor* or *area* or *symmetry* by hearing them used frequently in discussion as they investigate new concepts. This approach is compatible with current theories of second-language acquisition, which emphasize the use of new vocabulary in meaningful contexts while students are actively involved with objects, pictures, and physical movement.

Listed below are some key words used in this unit that will not be new to most English speakers at this level, but may be unfamiliar to students with limited English proficiency. You will want to spend additional time working on these words with your students who are learning English. If your students are working with a second-language teacher, you might enlist your colleague's aid in familiarizing students with these words, before and during this unit. In the classroom, look for opportunities for students to hear and use these words. Activities you can use to present the words are given in the appendix, Vocabulary Support for Second-Language Learners (p. 91).

giant steps, baby steps Students measure distances in the classroom by counting and comparing the number of *giant steps, baby steps* (heel-to-toe), and regular paces.

forward, backward, left, right, turn, robot These terms are used while students, working in pairs, take turns giving each other directions for moving a "robot" (the partner) from one point to another.

small, middle-sized, big Students use these terms to describe and compare the size of things they measure—their feet, adult feet, their paces, and classroom chairs.

middle, lowest, highest, in between Students measure their pace lengths and then try to determine the length of a standard pace by finding "the one in the *middle*." They do this while looking at their measures arranged in order of size, focusing on those *in between* the *lowest* and *highest* number.

In addition to these key words, students will encounter terms related to comparing their different "robot" routes (*straight path, route, distance*), to clothing (*clothes, sleeves, hat, pants*) and to rooms in their homes (*bedroom, kitchen, living room, furniture*). Familiarity with these words will be helpful as they work on the activities.

Multicultural Extensions for All Students

Whenever possible, encourage students to share words, objects, customs, or any aspects of daily life from their cultures and backgrounds that are relevant to the activities in this unit. For example:

- Students who have lived in other countries may be familiar with metric measure. When they explore metric clothing sizes at the measuring centers in Investigation 2, students may be able to bring in and share other examples of the use of metric measures, perhaps in sewing directions, sports and games, or recipes.
- Students will be measuring the distance around their heads as well as finding the lengths of their arms and legs. You might post a large sketch of a person and add labels for parts of the body in all the languages spoken in your classroom. This picture becomes a good resource for the class.
- In Investigation 4, students will be creating living spaces for little make-believe "Balobbies." You might explore cultural variations in the features of home and school interiors, and help students generate a list of words in their native languages that describe different rooms and the items in them.

Investigations

Measuring with Paces and Steps

What Happens

Session 1: Giant Steps and Baby Steps Students participate in familiar activities that link measurement to physical movement. They count their baby steps and giant steps as they measure distances in the classroom. As the students collect data and compare their results, they begin to notice and analyze the numerical differences that are produced when they use units of different sizes.

Session 2: Pacing and Comparing This session is devoted to comparing the number of units that are needed to cover a particular distance with giant steps, baby steps, and a walking pace. Students get firsthand experience with the inverse relationship between the size of their steps and the number of steps needed to cover a certain distance. They also estimate the number of paces needed to reach a target, then check their estimates.

Sessions 3 and 4: Robot Directions Students give directions to move a robot from one point to another, taking turns being the direction-giver or the robot. The directions involve both paces and turns. Learning to count off units (in this case, paces) along an unmarked distance provides students the opportunity to estimate and measure with familiar units.

Sessions 5 and 6: Finding the Middle-Sized Pace The "middle-sized pace" is proposed to students as a standard that will reduce the variation in their measurement of distances. The focus of these sessions is on finding the advantages of standardizing their paces and achieving more uniform results.

Mathematical Emphasis

- Using a nonstandard unit to measure a distance
- Experiencing the iterative nature of measurement (that is, counting the number of times an agreed-upon unit is used in moving from the beginning to the end of an object)
- Estimating length in "paces" by visualizing the unit "pace" repeated over a distance
- Comparing the effects of measurement using units of different size
- Describing the shape of the data
- Analyzing data by describing the patterns in the data
- Interpreting the meaning of certain landmarks in the data (e.g., "what is typical?")
- Examining a set of data to determine which is the "middle-sized" piece

What to Plan Ahead of Time

Materials

- Index cards or stick-on notes, for students to record data on (Sessions 1–6)
- Paper plates: 1 per pair (Sessions 3–4)
- Calculators: 1 per pair (Session 4)
- Adding machine tape, 1 or 2 inches wide: 2–3 rolls (Sessions 5–6)
- Scissors: 1 per pair (Sessions 5–6)

Other Preparation

- Introduce the game Giant Steps before beginning the investigation; see the **Teacher Note**, Playing the Giant Steps Game (p. 19). Encourage students to keep playing this game throughout the unit.
- Become familiar with making a line plot; see the **Teacher Note**, Line Plot: A Quick Way to Show the Shape of the Data (p. 20).
- Number or letter the paper plates, so that each pair can identify theirs. (Sessions 3–4)
- Duplicate student sheets, located at the end of this unit, in the following quantities:

  For Session 1

  Student Sheet 1, Giant Steps and Baby Steps: 1 per student (homework)

  Family letter (p. 95): 1 per student. Remember to sign it before copying.

  For Sessions 3–4

  Student Sheet 2, Pacing and Turning: 1 per pair

  Student Sheet 3, Pacing and Writing Directions at Home: 1 per student (homework)
- If you plan to provide folders in which students will save their work for the entire unit, prepare these for distribution during Session 1.

Session 1

Giant Steps and Baby Steps

Materials

- Index cards or stick-on notes
- Student Sheet 1 (1 per student)
- Family letter

What Happens

Students participate in familiar activities that link measurement to physical movement. They count their baby steps and giant steps as they measure distances in the classroom. As the students collect data and compare their results, they begin to notice and analyze the numerical differences that are produced when they use units of different sizes. Their work focuses on:

- estimating length
- using a nonstandard unit to measure distance
- comparing the effects of measurement using units of different sizes
- collecting and analyzing data

Activity

Estimating Distance in Giant Steps

This activity assumes that you have already introduced the game Giant Steps, either during class time or recess. You should also know how to record data on a line plot, as explained in the Teacher Note on page 20. Start class by reminding students about the game.

Remember when we played Giant Steps? Today we're going to estimate and then count distances in the room in giant steps.

Select a student to be the giant. The giant is to stand at the front of the classroom, then take two giant steps and freeze. Ask the other students to estimate the whole length of the classroom in those giant steps.

Make a picture in your head and try to imagine how many of Ricardo's giant steps it would take to get to the far wall.

Write the estimates on the board; then ask the giant to pace three more giant steps, and record any revision of the students' estimates. Next, as the giant paces the whole length of the room, have the students count out loud and record the answer.

You could each measure the length of the room in your own giant steps. Will all your answers be the same? What do you think?

Allow time for students to talk about whether they think there will be any variation in their results. Third graders have a wide range of theories about how and why measurement results might vary.

Measuring the Length of the Classroom Establish student pairs that will pace the length of the classroom in giant steps. While one partner paces, the other counts. Students then switch roles. Ask students to record their results on the board. They may look like this:

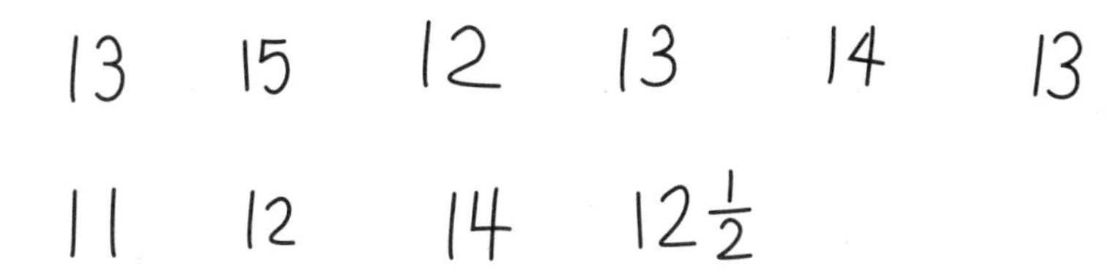

Look at the numbers on the board. Did everyone get the same results? Is that surprising? Why or why not?

Making Line Plots Make a line plot on the board to show the results of the pairs' pacing off the room. Draw a number line, label the points to include the smallest and largest numbers of giant steps, and show students how to make X's, check marks, or other symbols to record their data along the line.

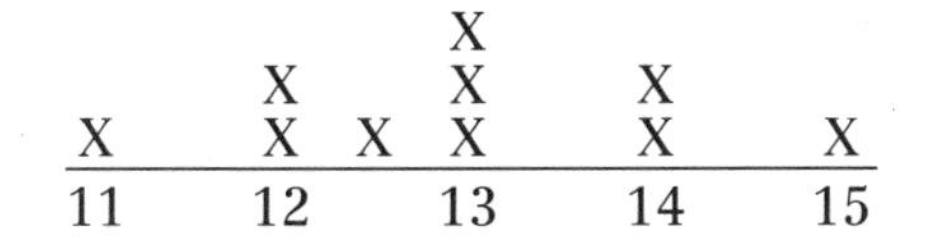

Why do you get different results? Can you think of some reasons?

Allow time for students to discuss the variation in results (which will surprise some students more than others). There is much room here for theory-building. For all the theories that are suggested, ask students for their reasons and evidence. Support their good thinking. Some of your students will have an easy time seeing that the size of a student's step makes a difference in numerical results; others won't. Inverse relationships—the smaller the pace, the bigger the number of paces—are sometimes hard to grasp, hold on to, and use.

Activity

Measuring in Baby Steps

Now you are going measure the same distance—the length of the room—but you'll use baby steps instead of giant steps. Do you think you'll get different results?

Allow a full discussion of students' predictions here, and when you think they're ready, ask someone to demonstrate baby steps.

Imagine, now, how many little steps like that it takes to get to the other wall. Try to imagine the baby steps in a straight path across the room. How many will it take?

Record students' predictions, then pair up students to pace the room in baby steps and count how many they take. When they have finished, have them write their results on index cards or stick-on notes, and then enter the data on a line plot.

Note: Save both the giant step and baby step data gathered in Session 1 for use in Session 2. Students will be comparing today's results with the results of their next class activity.

Session 1 Follow-Up

Homework

Student Sheet 1, Giant Steps and Baby Steps, encourages students to measure using giant steps or baby steps in their home environments. Send the family letter home with this sheet.

❖ **Tip for the Linguistically Diverse Classroom** For the second half of the activity on Student Sheet 1, give students the option of writing the starting and ending points in their native languages, or drawing simple sketches of where they began and ended.

Playing the Giant Steps Game

Teacher Note

This game (also called Mother, May I?) may be one your students know already. It usually involves 6–12 players. Optimal size is a group of about 10. You may want to divide your class and take turns playing—or if space allows, have two or three groups playing at once.

One player is the caller, who gives directions. The caller stands on an imaginary line facing the other players; they stand in a row about 15 to 20 feet away and move toward the caller as directed. The object of the game is to cross the line on which the caller stands. The first person over that line becomes the next caller.

The caller gives directions to each player in turn, such as "Take 5 giant steps" or "Take 7 baby steps," typically giving different instructions to each player. After hearing the directions, the player to whom they were directed must ask, "May I?" The caller then gives permission to perform the action with the words, "You may." Gradually, players advance toward the caller. A player who forgets to ask "May I?" must go back to the starting line.

In some versions of the game, the player proposes a move ("May I take 6 baby steps?"), which the caller can either allow ("Yes, you may") or disallow with a substitution ("No, but you may take 6 banana twirls"). The repertoire of possible moves includes:

- giant steps (the longest steps possible for the individual)
- baby steps (heel-to-toe steps)
- banana twirls (putting hand on head, stepping forward, and turning 360° simultaneously)
- bunny hops (hopping with both feet at once)
- lamp post (lying on the ground with feet marking current position—player reaches as far forward as possible, and stands at that point)
- duck steps (squatting, holding onto ankles, walking forward, quacking and flapping arms)
- anything else your students know (it will vary from one region to another)

Much of the game revolves around the caller imagining how many steps it will take to reach the finish line—and, of course, playing favorites about who will cross the line first to become the next caller. For your purposes, the point of the game is that students practice visualizing distances and use a variety of steps of different lengths.

As a variation, you might work specifically on visualizing measures of distance. Set yourself up as the caller and have two or three students stand in a row about 6 feet away. As you give directions (i.e., "Take 4 giant steps" or "Take 3 banana twirls"), ask the rest of the class to visualize whether the students will or will not reach you at the finish line with that move.

Teacher Note

Line Plot: A Quick Way to Show the Shape of the Data

An important part of statistics is organizing and representing data so that it is easy to see and describe. A *line plot* is one quick way to organize numerical data. It clearly shows the range of the data and how the data are distributed over that range. Line plots work especially well for numerical data with a small range.

A line plot is often used as a working graph during data analysis. As a working graph, it is an organizing tool we can use as we begin work with a data set, not a careful, formal picture we use to present the data to someone else. Therefore, it need not include a title, labels, or a vertical axis. A line plot can be simply a sketch showing the values of the data along a horizontal axis and X's to mark the frequency of those values in the data set. For example, if 15 students have just collected data on the number of paces it takes to walk the length of the classroom, a line plot showing these data might look like the one below.

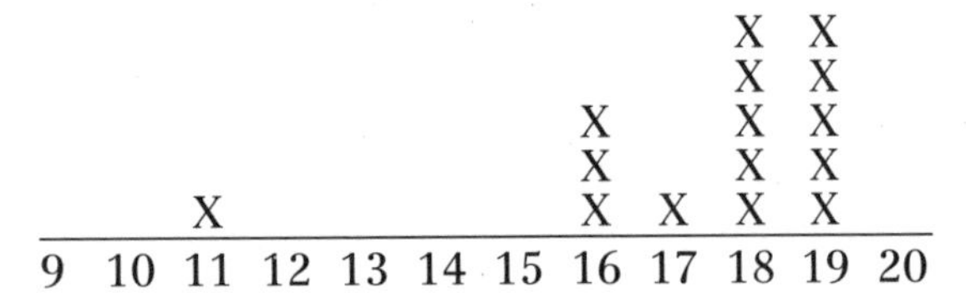

From this display, we can quickly see that two-thirds of the students took either 18 or 19 paces. Although the *range* is from 11 to 19, the *interval* in which most data falls is from 16 to 19. The *outlier*, at 11, appears to be an unusual value, separated by a considerable gap from the rest of the data. (These are terms that you will use with students as the need arises—introducing them informally in the context of discussing their data, rather than with formal definitions.)

One advantage of a line plot is that we can record each piece of data directly as we collect it. To set up a line plot, start with an initial guess from students about what the range of the data is likely to be: What do you think the lowest number should be? How high should we go? Leave some room on each end of the line plot so that you can lengthen the line later if the range includes lower or higher values than you expected.

By quickly sketching data in line plots on the chalkboard, you provide a model of how such plots can provide a quick, clear picture of the shape of the data.

Session 2

Pacing and Comparing

What Happens

This session is devoted to comparing the number of units that are needed to cover a particular distance with giant steps, baby steps, and a walking pace. Students get firsthand experience with the inverse relationship between the size of their steps and the number of steps needed to cover a certain distance. They also estimate the number of paces needed to reach a target, then check their estimates. Their work focuses on:

- using a nonstandard unit to measure distance
- comparing the effects of measurement using units of different sizes
- estimating distances
- collecting and analyzing data

Materials

- Index cards or stick-on notes

Ten-Minute Math: Estimation and Number Sense Once or twice during the next few days, use the activity Estimation and Number Sense. Remember, this short activity need not be done during math time.

Display an addition problem on the board for about a minute. For example:

25 + 10 + 25 or 31 + 46 + 3

During this minute, students mentally estimate the answer—no writing or calculator use allowed.

Cover the problem while students discuss their estimates.

Reveal the problem again and ask them to find a precise solution to the problem by using mental computation strategies.

The first two times you do this activity, use only addition problems. For variations, see pp. 87–88.

Activity

Pacing the Classroom

Recently we measured the length of our classroom in giant steps and baby steps. Taking giant steps or baby steps can be uncomfortable, though. A more widely used measurement is the *pace*, which is closer to a regular step. How can we measure someone's pace? Let's try with a volunteer.

Ask one student to take three or four regular walking steps, then freeze. Ask the class how long they think the room is in paces of that size. After they estimate the distance, have the same student pace off the length of the room while the class counts to keep track.

Working in the same pairs as in Session 1, the students estimate and count the number of paces it will take them to measure the length of the classroom. You will record their data on a line plot, as before.

Let's record our results on a line plot. We need to decide on the beginning and end of the number line we'll use. Any ideas?

What can we see from this line plot? What can you say about the data?

Help students express their ideas as they describe the distribution of the data (see the **Teacher Note**, The Shape of the Data: Clumps, Bumps, and Holes, p. 25).

What do you see in the graph? Do you see any clusters of data? Did most of you get the same number of paces when you measured the length of the room? Is the range of data very wide? Are there any unusual values?

For an example of such a discussion, see the **Dialogue Box**, Describing the Shape of the Data (p. 24).

On another line plot, put the results of Session 1 (giant steps and baby steps) on the board and ask students to compare those data with their new results.

What can we say about our paces, our giant steps, and our baby steps? How do they compare?

This is likely to produce a discussion of the relationship between step size and number of steps. Students will argue with each other about this relationship and about reasons for the difference in results. They may have trouble articulating their ideas, and some will not be able to put their ideas into words. Students may assert that it takes more steps for bigger people; others will assert that it's fewer—again, the inverse relationship (the bigger the person the fewer the steps) may confuse some students. Some teachers who have worked with this unit believe that the fact that smaller numbers "win" here is unusual and important, so that students do not just look at large numbers without evaluating what they represent.

(See the **Dialogue Box**, Discovering Children's Beliefs About Numbers, p. 26, for another example of how classroom discussions can help you discover how your students think about number, measurement, and data.)

Questions like the following can produce fruitful discussion:

Which are the biggest steps? Did you take more baby steps or more giant steps? Why?

If I take 6 giant steps to walk along here, will I take about 30, or about 10, or more like 3 baby steps to cover the same distance?

Make sure that students give reasons for their statements.

Activity

Estimating How Far in Paces

Ask one student to stand in a fairly open part of the classroom. Select a target that is a moderate distance away *in a straight line*. Have the student take three or four *paces* to help the others visualize the length of a pace.

You may want to dramatize the visualization process by "thinking aloud" your own way of making an estimate:

Let's see—I can see how long Chantelle's pace is, so I'll try to imagine: 2 ... 3 ... 4 ... 5 ... 6. About 6 paces to the desk, I think. Let's try it!

Now, how many paces is it from Chantelle to the globe?

Students estimate, the pacer paces the distance, and everyone counts. Repeat this two or three times, selecting different objects. You may want to use more than one student as a pacer.

Session 2 Follow-Up

Estimating Other Distances Students may develop some good strategies for estimating distances in paces. To practice these strategies, follow up with a whole-class activity. Select various points in the classroom and ask everyone to estimate the number of a specific student's paces between the points. Write down all estimates, then check by pacing to see how close the class estimates are. Visual estimation is an important component of measurement, and practice does make a difference!

DIALOGUE BOX

Describing the Shape of the Data

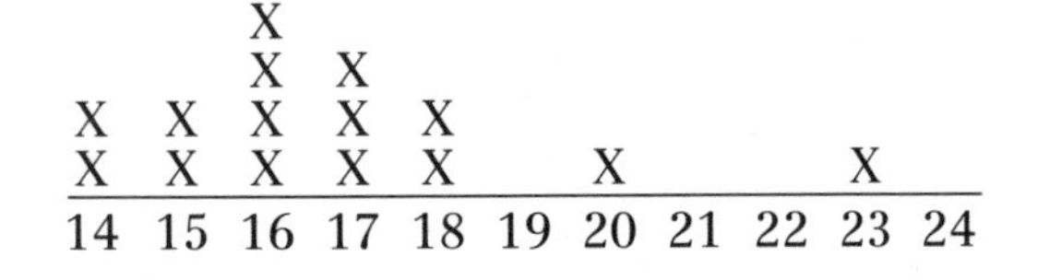

The students have just counted the number of paces it takes each of them to walk the length of the classroom. They have recorded their data on a line plot (above).

So what can you say about these pace data? Let's hear a few of your ideas.

Sean: Well, there are a lot at 16.

Chantelle: There's only one at 20 and at 23.

Kate: There are two each at 14 and 15.

What else do you notice?

Maya: Fourteen is the lowest.

Jeremy: Yeah. And 23 was the highest.

So the *range* was from 14 to 23. What else?

Annie: There's nothing at 19, 21, or 22.

Annie's noticing a lot of holes in this part of the data. Can anyone say any more about that?

Michael: Well, there's nothing at 12 or 13 either.

Yes, 14 is the lowest count and there's nothing below it. But this situation, that Annie noticed up here, is a little different. What can you say about that?

Su-Mei: Mostly, the paces go from 14 to 18, but sometimes you get something higher.

Can anyone add to that?

Sean: You must have really small paces if it takes you 23.

In fact, mathematicians have a name for a piece of data that is far away from all the rest. They call it an *outlier*. An outlier is an unusual piece of data—sometimes it might actually be an error, but sometimes it's just an unusual piece of data. It's usually interesting to try to find out more about an outlier. Who had the outlier in this case?

Ly Dinh: I did. And I counted twice, and Michael checked me, too, so I know it was 23 paces.

Jennifer: Maybe he's got smaller feet.

Any other theories about Ly Dinh's pace?

[*Later*] ... **Suppose someone asked you, then, "What's the typical number of paces to cross our classroom?" What would you say?**

Tamara: Well, I'd say 16.

[*Addressing the class as a whole, not just Tamara*] **Would 16 be a reasonable description of how many paces long our room is?**

Cesar: Yes, because most of us took 16 paces.

Any other ways to say this? Or any different ideas?

Midori: Well, I wouldn't say just 16.

Why not?

Midori: Well, there's really not that much difference between 16, 17, and 18. They're all really close together. I'd say 14 to 18, 'cause the 20 and 23 aren't what you'd usually get.

So Midori is saying she'd use an *interval* to describe the pace-length of our room, from 14 to 18, and Cesar said he'd say the length was about 16. What do other people think?

In this discussion, the class has moved gradually from describing individual features of the data to looking at the shape of the data as a whole. Throughout, the teacher tries to have students give reasons for their ideas and pushes them to think further by asking for additions or alternatives to ideas students have raised.

Notice how the teacher introduced, in context, the terms *interval*, *range*, and *outlier*. This is the best way to introduce such new vocabulary, when the ideas come up and the terms describe some particular real data.

The Shape of the Data: Clumps, Bumps, and Holes

Teacher Note

Describing and interpreting data is a skill that must be acquired. Too often, students simply read numbers or other information from a graph or table without any interpretation or understanding. It is easy for students to notice only isolated bits of information (e.g., "Vanilla got the most votes," "Five people were 50 inches tall") without developing any overall sense of what the graph shows. Looking at individual numbers in a data set without looking for patterns and trends is something like decoding the individual words in a sentence without comprehending the meaning of the sentence.

To help students pay attention to the shape of the data—the patterns and special features of the data—we have found it useful to use such words as *clumps, clusters, bumps, gaps, holes, spread out, bunched together,* and so forth.

Encourage students to use this casual language about shape to describe where most of the data are, where there are no data, and where there are isolated pieces of data.

A discussion of the shape of the data often begins with identifying the special features of the data: Where are the clumps or clusters, the gaps, the outliers? Are the data spread out, or are lots of the data clustered around a few values? Later on students decide how to interpret the shape of these data: Do we have theories or experience that might account for how the data are distributed?

By focusing on the broad picture, the shape of the data, we discover what those data have to tell us about the world.

DIALOGUE BOX

Discovering Children's Beliefs About Numbers

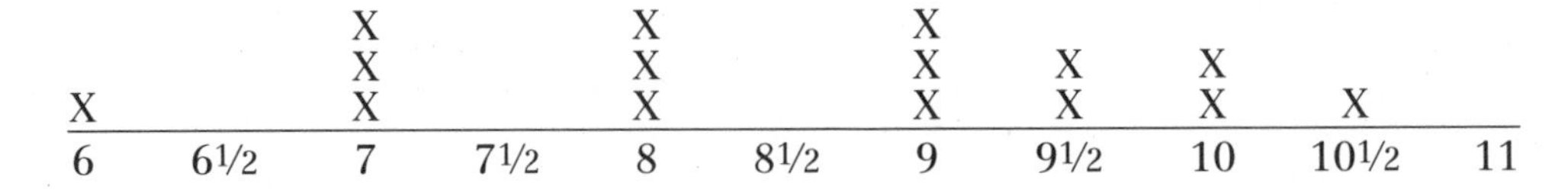

Children have many ideas about numbers that are not always apparent in the classroom. One of the benefits of mathematical discussions is that those ideas can surface and be talked about. Here, for instance, is a discussion that occurred in a third grade classroom while students were analyzing data about foot length. The display (above) showed the lengths of their feet in inches, and they were looking for the middle-sized foot. (Your students will do this activity themselves in Investigation 2, Session 2.)

Amanda: I think 8 is in the middle because 9 1/2 and 10 1/2 aren't numbers, they're halves. Then if you match 6 and 10, and 7 and 9, 8 is in the middle.

Tamara: I think you should count 9 1/2 and 10 1/2 too, because they're numbers.

Ricardo: I think so, too.

Amanda: I still think that 1/2 numbers aren't regular numbers.

Elena: But we have to count those numbers because those numbers have people. We don't need to count the other halves because nobody has those numbers.

Amanda: I get it. I think we should count them too, but only if they have people.

The discussion clearly had two threads—the "reality" of the fractional numbers, and the search for the middle-sized foot. The students resolved it by granting reality to the halves when they represented real values (that is, if they "had people").

The teacher did not take an active role in this discussion, preferring that the students talk among themselves about whether the numbers were real or not. She was surprised by their idea that halves weren't numbers, and decided to follow up in a different context at a later date.

Gaps between children's intuitive, informal understanding of numbers and their school knowledge do not always arise unless there is some reason to talk in a fairly open-ended way about numbers used in a real context. One of the important benefits of mathematical discussions is the chance to hear students expressing their intuitive beliefs about numbers. Informal diagnosis of students' understanding is often based on such conversations.

Sessions 3 and 4

Robot Directions

What Happens

Students give directions to move a robot from one point to another, taking turns being the direction-giver or the robot. The directions involve both paces and turns. Learning to count off units (in this case, paces) along an unmarked distance provides students the opportunity to estimate and measure with familiar units. Their work focuses on:

- estimating distance to various points
- giving oral, then written directions that involve nonstandard measurement of distance
- comparing lengths of routes

Materials

- Paper plates (1 per pair, as targets)
- Student Sheet 2 (1 per pair)
- Student Sheet 3 (1 per student, homework)
- Calculators

Activity

Giving Directions to Reach a Target

If someone asks you for directions to another place, it's important to give them good directions. Today you will pretend to be robots and practice giving directions by telling each other how far to move to get to a target.

Demonstrate the task in front of the class. Ask for a volunteer robot, who must do exactly what you say. Put a paper plate some distance away from the robot, directly in its path; this is the target. Give directions that will move the robot to the target paper plate. Use only "forward" or "backward" commands; you will add turns in the next activity. Say, for example:

Robot, go 5 paces forward.

Assess the effect of this movement with the class.

Does the robot need to be given new directions? Did my estimate as the direction-giver work for this particular robot?

Students work in pairs, taking turns being the robot. Give each team a paper plate identified by a number or letter. The direction-giver positions the plate, tells the robot how many paces to take toward the target, and assesses the success of his or her directions. Then students swap roles. They do this twice, then return to their seats.

Was this difficult? Were your estimates of distance fairly accurate?

Encourage students to talk about their experiences, calling attention to the issue of estimating distance in paces.

Activity

Pacing and Turning

Tell students that now you're going to be giving more realistic directions that involve both pacing and turning.

We'll practice making turns and measuring distances that aren't in a straight line. First, I need another volunteer robot.

This time, write simple directions on the board for the robot that involve making turns.

Turn left. Go 4 paces forward. Turn right.

Help students establish a working definition of a turn. Most classes decide that a turn is a 90° turn, or a square corner. Point out a target in the classroom and give directions to the robot as a demonstration. The robot is to move as directed after each command.

Explain to students that this time when they give directions, they'll be writing them down so that they have a record of them. Distribute Student Sheet 2, Pacing and Turning, to each team.

Now you'll be writing directions that involve turning and pacing. When you write, you'll be writing down the same thing you would say to the robot. Don't get too fancy—just give the robot three or four commands to reach the target.

❖ **Tip for the Linguistically Diverse Classroom** If you have students who are not yet reading and writing in English, encourage them to devise pictures and symbols or to use their native language to convey their directions. Thus, on Student Sheet 2, directions could be written entirely nonverbally. You could also pair students who are proficient in English with those who have limited English proficiency.

For the last part of the session, students again work in pairs, taking turns being the robot. They place their target, then write down a few simple directions to reach the target, such as "Forward 3 paces. Turn left. Forward 2 paces." Each time they are the direction-giver, they choose a new target location and fill in a box of robot directions on Student Sheet 2.

This will not be easy for students! Estimating how many paces and what kind of turns are needed to reach the target is a complex task. Students will need more practice with this, and will continue working on it during the next session. The homework will also give them experience with giving directions.

Teacher Checkpoint

Do Students' Directions Work?

Spend about half of Session 3 having students continue their pacing and taking turns writing directions for reaching a series of different targets. While they work on Student Sheet 2, circulate around the room and check in with students to see how their directions work.

With each pair, choose one set of directions and read them as commands to the "robot" that the team has identified. See if each team's directions work to reach the target.

- If the directions that are written do not work well, ask the pair to discuss what part of the directions went wrong. Ask them how could they change the directions so they work better.
- Where are students having the most trouble: with accurate specification of the number of paces, with direction of turns, or simply with being clear in their communication? Addressing this question will help you figure out where they need more practice.

See the **Teacher Note**, Connecting Oral and Written Directions (p. 31), for ideas about further experiences to help students with writing directions.

Activity

Planning and Comparing Robot Routes

Now you're all going to write robot directions from one place in the school to another. When you're finished, we'll compare your routes.

Help students decide, as a class, on a starting point and an ending point for the robot. It's more interesting to do this activity if you aren't limited to your own classroom. For example, you might try moving the robot from the door of your classroom to the library, or the office, or the front door. Write the starting and ending points on the board and be sure each pair records them before devising their route. Encourage students to think of different ways of getting from the starting to the ending point.

Students then work in pairs to figure out their own route. Pairs may need to take turns working out their routes so that they don't get in each other's way. Once the route is set, they add up the total number of paces in their route. Let them use calculators to check and compare the distances of their routes.

Finding the Shortest Route When student pairs have established their routes, gather as a whole group to collect their data.

We've found some different routes to get to the same place! Let's look at our routes between [our classroom door] and [the library]. We'll compare all the routes to see whose is the shortest. Who wants to read their route aloud?

As each pair reads their route, record it in a large grid on the board or on the overhead projector so that the whole class can see it.

TEAM	ROUTE							TOTAL PACES
Elena and Tamara	Forward 2	Right	Forward 7	Left	Forward 9	Right	Forward 5	23
Yoshi and Kate	Forward 4	Right	Forward 2	Left	Forward 4	Right	Forward 10	20
Ly Dinh and Jeremy	Forward 6	Right	Forward 3	Left	Forward 6	Right	Forward 10	25

Whose route is longest? Whose is the shortest? Is there something special about the shortest route of all?

Follow the same procedure to find the longest route in class. Compare and discuss these routes.

We've spent time finding the very shortest route and the very longest route anyone wrote down. Can you find an even longer one? Yes, Ryan, come up and show us. Khanh, you come too....
Can anyone think of an even longer one? Talk with your partner and together write down the very longest route from [the class door] to [the library] that you can think of.

You'll find that the students can generate even longer routes once they start to have fun with the task. Just as you can always find a larger number by adding one more, you can always find a longer route by adding some turns or some backward steps.

Students' strategies for getting the longest distance involve overshooting the target and turning a lot, and will produce a good laugh for the whole class. This part of the session should result in some shared enjoyment, but more important, it will give the students a sense that they know how to give directions, with turns, using measurements to describe distance.

This activity is a good place to use calculators; have one or two students pace a route while the others add up the paces they've taken.

Session 3 and 4 Follow-Up

Session 3 Each student will need a copy of Student Sheet 3, Pacing and Writing Directions at Home. Students pace off distances at home, as they did in class, to practice their estimation and measurement skills. The homework gives them a chance to write down directions to various targets, and to choose targets of varying degrees of complexity.

Using the Computer and Logo If you and your students have been using Logo on the computer, working with this software language is a logical extension and reinforcement of these activities. Giving and responding to directions about turns and distances is similar to moving the Logo turtle around the screen. You can organize the task so that one student points to a place on the screen and the other moves or programs the turtle to get to that spot. You can save the pictures or the programs and have students share and compare their solutions.

Connecting Oral and Written Directions

Teacher Note

When students give oral directions and commands to a "robot," they can see the effect of their words immediately. Creating a series of written directions, however, involves taking on another person's orientation in space. Telling the robot whether to turn right or left when the robot is facing in a different direction is very difficult.

Sometimes students need time to realize that the directions they write down are just the same as those they would have said out loud. It may help them to think through the effects of one move at a time, jotting down the directions as they think them. Sometimes students may need to write out and then check their directions.

Orientation in space is a complicated issue for students of this age. Some are better able than others to visualize the movement of someone else in space. Many students will need more experience before they can visualize turning left and right from another's point of view. These variations in ability are typical for third and fourth graders. Working carefully through this investigation and listening to other students describe their visualization methods will help students who are more hesitant.

Sessions 5 and 6

Finding the Middle-Sized Pace

Materials

- Adding machine tape (2–3 rolls)
- Scissors

What Happens

The "middle-sized pace" is proposed to students as a standard that will reduce the variation in their measurement of distances. The emphasis of these sessions is on finding the advantages of standardizing their paces and achieving more uniform results. Their work focuses on:

- finding a standard of measure
- displaying data and finding the middle

Ten-Minute Math: Estimation and Number Sense Once or twice during the next few days, continue to use the activity Estimation and Number Sense as described in Session 2 (p. 21). Remember to find time for it outside the math hour. This time, present addition and subtraction problems that require reordering; for example:

$6 + 2 - 4 + 1 - 5 + 4 + 5 - 2$

$36 + 22 + 4 + 8$

Display each problem for about a minute, while students work out their mental estimates.

Cover the problem and ask students for their estimates.

Uncover the problem and ask students for a precise solution, using *mental* computation strategies (no paper and pencil or calculator).

You may also want to try money problems; for example:

$1.25 + $2.63

$$\begin{array}{r} \$5.13 \\ \$6.50 \\ \underline{\$3.30} \end{array}$$

For other variations, see pp. 87–88.

Activity

What's the Middle-Sized Pace?

When we talked about giving directions to robots, a number of you wanted to find one pace for everyone to use when they walked, so that our directions would always come out the same.

Today we will look at everyone's pace and try to find the middle-sized pace to use as a standard. That way you'll all be able to give similar directions. How could we go about finding the middle-sized pace?

Listen to all their ideas about solving this problem. List ideas on the board as they come up so that students can develop and modify each others' plans. This should be considered an introductory discussion to get them thinking about the problem; they need not settle on a plan at this time.

Some of your students will suggest that you standardize these measurements by using inches or feet. Others, though, won't even think about standard units of measurement until later in this unit. For those students, it's important to hold off on the standard units of measure for a while. If someone suggests inches or feet, you might respond, "That's a good idea, Jeremy, but let's wait. We'll use them later."

Measuring Paces After the opening discussion, turn to the issue of measuring a pace.

How could we find the length of everyone's pace? Let's start out by finding how long my pace is. How could we use this paper tape?

Solicit ideas for measuring your pace directly. Take two or three paces and freeze. Have two student volunteers mark your pace on the tape and use scissors to cut it to size, following directions from the class.

There is lots of opportunity for discussion here, and students need to agree on a single method before the volunteers cut the tape. Will they measure toe to toe or heel to heel? (A surprising number want to measure toe to heel!) Be sure that they give reasons for choosing a particular method. You may need to dramatize the differences between suggested methods by having the volunteers cut many different tapes for your pace, following the different methods, and then compare them.

Once the class has settled on a method, establish working groups of two or three and give each group a pair of scissors, a pencil, and a length of adding machine tape. Helping one another, students in a group cut the paper tape to the length of each individual's pace. Writing names on the pace-tapes will help later when students compare sizes.

Looking at Everyone's Pace As the groups finish, fasten their pace-tapes randomly to the wall or chalkboard. Ask students how they propose to find the middle-sized pace. Some may suggest ordering the tapes by size (smallest to largest, or vice versa); try not to accept that suggestion too quickly. Wait to hear many ideas, and ask students to demonstrate or explain so that others understand their methods. When the discussion has run its course, enlist volunteer help in ordering the pace-tapes by size.

Now that we have all our paces, how could we find the middle-sized pace?

Students will have some good ideas about finding the middle of a set of data. Probe their ideas and question them so that their reasoning becomes clear. For two different directions this discussion might take, see the **Dialogue Box**, Finding the Middle-Sized Pace (p. 37).

Elena says to look for the "middle." But how could we do that? Which one is in the middle? How can you tell? Can you be sure? What is the middle-sized pace?

After your class has agreed on a method, identify the middle-sized pace. Allow some time to talk about any surprises. Is the middle-sized pace very different from the others? Are the pace-lengths clumped?

Activity

Teacher Checkpoint

Comparing Paces to the Standard

Hold up a copy of the middle-sized pace. Ask individual students to come up and compare their pace to the middle-sized pace. Try to choose students who have a pace shorter than, longer than, and the same size as the pace you are holding.

Let's suppose that we used this middle-sized pace to measure a distance and we found that it took 7 of these paces. If you measured that same distance using your pace, how do you think the total number of paces would compare?

As students offer explanations, listen for evidence that supports the idea that different results come from using somewhat different units and that the smaller the unit the greater the number needed to measure, and conversely, the larger the unit the smaller the number needed to cover the distance.

If students have difficulty comparing their total number of paces to the total number of middle-sized paces, they may need more experiences with measuring with nonstandard units and comparing totals, as provided in Sessions 1–4.

Activity

Using the Standard Pace

Now that we have agreed on the middle-sized pace, let's make copies of it with adding machine tape. Then all of you can use paces of that size to measure distances in the classroom. If we measure the length of the room with the middle-sized pace, will our results vary? Let's try it.

Students may want to mark the middle-sized pace on the floor so they can practice standardizing their own walking paces. Keeping copies of the standard on the paper tape will allow students to settle debates by referring to the tape itself. In this way the middle-sized pace will truly function as a standard measure for your class.

Set some other measuring tasks that involve pacing distances with the new standard.

Let's see how far it is to the school office in standard paces, and how far to the bathrooms. At lunch time, see how many standard paces it takes to get to the cafeteria.

Sessions 5 and 6 Follow-Up

Extension

Measuring with the Standard Students can use the standard pace-tape as a ruler to measure objects in the classroom. They might measure things such as the teacher's desk, the chalk tray, the bulletin board, the windows—even the heights of their classmates.

DIALOGUE BOX

Finding the Middle-Sized Pace

The discussion in the activity What's the Middle-Sized Pace? (pp. 33–34) varies from class to class. Following are examples from two different classrooms. The first is a conversation that happened after students had paced to predetermined targets in the school. They found that while the directions they wrote down worked for the one who originally paced the distance, they did not always work for the other partner. The teacher asked the students about that experience.

How many of you had the problem that your partner's paces weren't the same size as yours, so the number didn't come out right? What do you think we could do about that?

Tamara: You can measure our paces, measure all of them, and find the biggest and the smallest and then you get the middle one and try and make everybody take that size of a pace.

Jamal: The ones with the middle-sized pace come to the front of the room and walk. Then we can see who has the middle.

Dylan: Rearrange it, the tall ones on one end, and it goes down to the other ones that way.

Samir: Yeah, that sounds neat! We'd organize it so the smallest one would be first, then the next smallest one ... all the way up to the biggest one. They'd be all in a row like that.

How would that help us find the middle size?

Tamara: Because the smallest one, you can start from there and go until you get to the one in the middle, and that could be the middle size.

Any other ideas to help us find the middle-sized pace?

Liliana: I was going to measure them and then try to find the middle.

So actually write down how long the longest one was and how long the shortest one was and all the ones in between?

In a second classroom, the students did not want to select one pace as the middle one, but instead kept selecting a range of values as their middle. Because this was so important to them, the teacher did not push for the selection of one middle-sized pace.

During this discussion, the students were looking at a line plot that showed how many paces each of them took to cross the room.

				X				
				X		X		
X	X	X		X		X	X	
X	X	X	X	X	X	X	X	X
12	13	14	15	16	17	18	19	20

Ari: These paces are the middle ones. [*Marks under 14, 15, and 16 on the chalkboard.*]

Is there one pace that's in the middle? There are 18 paces in all. Can we use that to find the middle?

Sarah: This is the middle one. [*She points to 16.*]

Kevin: Here. 1, 2, 3, 4, 5, 6 ... these are the big paces [*indicates 12 to 14*] and 1, 2, 3, 4, 5, 6 ... these are the small ones [*indicates 18 to 20*] and 1, 2, 3, 4, 5, 6 ... these are the middle ones [*indicates 15 to 17*], all six of them here.

What if we wanted to tell someone how long our middle-sized pace was?

Marieke: Just tell them in between our small-size ones and the big-size ones.

Sarah: In between here and here. [*Points to 14 and 18.*]

Can we be more specific?

Jonathan: Say it's 17½ inches, because it is. It's mine and I measured to be sure.

How do you know yours is the middle?

Jonathan: It's in the middle group.

Sarah: But mine is too and it's shorter than yours.

From Paces to Feet

What Happens

Session 1: The Need for a Standard Measure Students demonstrate their understanding of the value of standard measures as they respond to a story about a king who has difficulty communicating building directions to a carpenter. They are introduced to standard measuring tools and use rulers to collect data about the size of their feet.

Session 2: Kids' Feet and Adults' Feet Students use sketch graphs and line plots to organize and analyze the foot data they have collected about the size of their feet and the size of adult feet. They use rulers to measure an object larger than 12 inches and discuss their ways of measuring.

Sessions 3 and 4: Measuring Centers In four activities organized in measuring centers around the classroom, students use standard measuring tools to measure familiar objects in the classroom; they locate "benchmarks" on their body, which they can use to estimate lengths without a ruler; and they collect and discuss data about how far they can jump and how far they can blow a pattern block.

Session 5: Moving to Metric Students individually analyze the data they have collected at the pattern block measuring center, allowing you to assess their abilities in data analysis. They are introduced to centimeters and meters as a different system of measurement used by most of the world, and they begin to develop centimeter awareness by making their own centimeter and meter measuring tapes. They then use these tools on a scavenger hunt in their classroom, looking for things that are about 1 meter and about 1 centimeter long. For homework, they continue the scavenger hunt in their own homes.

Sessions 6 and 7: Metric Measurement Students conclude their scavenger hunt by examining the data from home and school. Then they make metric measurements related to their clothing sizes. One of these measurements—head size—is used to assess how well students can measure using centimeters. Throughout the sessions, they learn more about meters and centimeters and how to use them.

Mathematical Emphasis

- Understanding the rationale for a standard measure (that is, the need for consistency and accuracy)
- Developing familiarity with inches, feet, and yards—what these units are, and when and how to use them
- Developing awareness of centimeters and meters and how big these units of measure are
- Learning to describe a set of data that involve measurements, first representing these data on a line plot and then describing the general features of the data set

What to Plan Ahead of Time

Materials

- *How Big Is a Foot?* by Rolf Myller (Dell, 1990) (Session 1, optional)
- Measuring tools:
  Tape measure: 1 for display (Session 1)
  Inchsticks: 1 per pair (Sessions 1–4)
  Rulers marked in inches and centimeters: 1 per pair (Sessions 1–7)
  Yardsticks: 4–5 to share (Sessions 3–4)
  Centimeter cubes (Session 5) (optional)
  Metersticks: 4–5 to share (Sessions 5–7)

  Note: Please refer to A Note on Measuring Tools (p. 11) for important information on dealing with combination yardsticks and metersticks.
- Stick-on notes or index cards: 1 per student (Session 1)
- Pattern blocks or checkers (Sessions 3–4)
- Chalk (Sessions 3–4)
- Tape (Sessions 3–5)
- World map or globe (Session 5)
- Scissors (Sessions 5–7)
- Yarn or string: allow about 60 cm per student (Session 6–7)
- Chart paper: about 12 sheets (Sessions 3–7)
- Blank overhead transparencies: 3–4, for making line plots (you may use the board or chart paper if you prefer)
- Overhead projector (if you use transparencies)

Other Preparation

- Duplicate student sheets and teaching resources, located at the end of this unit, in the following quantities:

  For Session 1

  Student Sheet 4, The King's Foot: 1 per student

  For Sessions 3–4

  Student Sheet 5, Measuring Center Data Sheet: 1 per student

  Student Sheet 6, Measure and Compare: 1 per student

  For Session 5

  Student Sheet 7, Metric Scavenger Hunt: 2 per student

  One-centimeter graph paper (p. 112): 1 sheet per student

  For Sessions 6–7

  Student Sheet 8, My Sizes in Metric: 1 per student
- If you don't have inchsticks, make at least 1 per pair from the pattern provided (p. 111).
- As needed, make transparencies of Quick Image Dot Patterns (p. 113) and Geometric Designs (p. 114) and cut apart the images for the Ten-Minute Math activities.
- Be aware that for Sessions 3–4 and Sessions 6–7, you will need to set up measuring centers in the classroom. Specific instructions are given in the session activities.

Session 1

The Need for a Standard Measure

Materials

- *How Big Is a Foot?* by Rolf Myller (optional)
- Student Sheet 4 (1 per student)
- Rulers or inchsticks (1 per pair)
- Yardstick, tape measure (1 each for display)
- Stick-on notes or index cards (1 per student)

What Happens

Students demonstrate their understanding of the value of standard measures as they respond to a story about a king who has difficulty communicating building directions to a carpenter. They are introduced to standard measuring tools and use rulers to collect data about the size of their feet. Their work focuses on:

- communicating ideas about the need for a standard unit of measure
- using a ruler as a standard measuring tool
- collecting data through measuring

Activity

Assessment

The King's Foot

"The King's Foot" is a story about a king, a carpenter, a horse stall, and the importance of agreeing on a standard unit of measure.

Note: *How Big Is a Foot?* has a similar theme. In this book, the carpenter's apprentice gets thrown in jail because the bed he makes for the queen is too small. If you can get this book, you can substitute it for "The King's Foot." At the point in the book when the apprentice is thrown in jail and the story asks, "Why was the bed too small for the queen?" pause so that students can respond in writing to this question, composing a letter to the carpenter's apprentice. See Introducing the Assessment Task, p. 43, for more explanation. This assessment was suggested by Marilyn Burns in *Math and Literature, K–3* (Math Solutions Publications, 1992), for use with *How Big Is a Foot?*

Introduce either story by relating it to the students' measuring experiences in this unit.

Earlier, we measured distances—like the length of our classroom, and how far it is from our room to the front door of the school—by pacing them off. We also tried using giant steps and baby steps. Here's a story about a king who measured the same way. Listen to find out how well it worked for him.

Read the story to your students. They might act out the story as you read it.

❖ **Tip for the Linguistically Diverse Classroom** If you are using the book *How Big Is a Foot?*, be sure to show the pictures as you read it. If you are reading "The King's Foot" aloud, having students enact the story will help make it comprehensible to students with limited English proficiency. Simple props (a king's crown, a yarn mane for the pony, a carpenter's hammer, an area marked off for the horse's stall) will also help.

The King's Foot

Once upon a time there was a king who kept ponies. His daughter, the princess, had a little pony of her own that she dearly loved. As the princess grew older she grew bigger, but the pony did not. The day came when she climbed on her pony and her feet dragged on the ground. That was the day the king decided that he would surprise his daughter with a beautiful new full-size horse.

The king went to the best stable in the kingdom and chose a sleek Arabian mare. "Because it's a surprise," the king said, "I want to leave the mare here at your stable until I can get a new stall built in the royal barns to fit such a grand, large horse."

The king knew that he would have to tell the royal carpenter how large to make the stall. So, using heel-to-toe baby steps, the king carefully walked around the mare, imagining how big the stall for this beautiful horse should be.

"... 5, 6, 7, 8, 9 feet long," he murmured, "and 3, 4, 5 feet wide. I will tell the royal carpenter to build a stall that is 9 feet long and 5 feet wide."

Continued on next page

The king jotted down the numbers: 9 feet long and 5 feet wide. The message was sent to the carpenter, and she set to work at once.

Soon the stall was ready and the king sent for the mare. He thought he would have a little fun with the princess, so he had the royal groom hide the mare behind the barn. Then he said to the princess, "Come with me and see if you can guess your surprise."

Together they walked into the royal barn, past all the stalls of little ponies. They stopped in front of the empty new stall. But no sooner had the princess inspected the new stall than she burst into tears.

"I truly hoped that my surprise would be a horse, because I have outgrown my little pony. But now that I see the size of the stall, I know that you are just giving me another little pony, no larger than the first."

The king was puzzled. He saw that indeed, the new stall was much too small for a full-size horse. The groom quickly brought the new Arabian mare out of hiding, and as soon as the princess laid eyes on her, she forgot her tears. Only the king did not forget. He called angrily for the royal carpenter to account for her terrible mistake.

The carpenter was shocked. She knew she was good at her trade; her work always drew high praise. And she had made the stall just as the king had said—9 feet long and 5 feet wide. She had been very careful to use heel-to-toe baby steps, 9 feet long and 5 feet wide, when she measured the size of the stall. What could have happened?

[***At this point in the story, pause so that students can do the assessment task. See Introducing the Assessment Task (p. 43).***]

The carpenter stared sadly at her work. She paced thoughtfully around the little stall, carefully counting her foot-lengths. Then she sat down beside the king to think, staring at her feet.

That was when the carpenter noticed something—when she saw the king's foot next to hers. "That's it!" she cried. "Your foot is much longer than mine! I made the stall 9 feet long, but I used 9 of my feet instead of 9 king's feet."

Then the carpenter had a truly remarkable idea. She took a flat stick of wood, and she cut it just exactly the same length as the king's foot. "This way," she told the king, "I can always know exactly how big you want things made."

Now the carpenter made a stall for the new horse that was 9 king's feet long and 5 king's feet wide. This time the stall fit perfectly. So the king was happy, and the princess was happy, and the carpenter was happiest of all. She started a factory and made lots of sticks just as long as the king's foot, which she called rulers. Selling these sticks, she became rich and famous.

THE END

Introducing the Assessment Task Distribute Student Sheet 4, The King's Foot, to each student. (You will need to adapt this sheet if you are using *How Big Is a Foot?*)

The story is asking us what could have happened. Why was the stall that the carpenter built too small for the new horse? Before we finish reading the rest of the story, write a letter to the carpenter explaining what happened. Your letter should include an explanation of what went wrong, what she could do to correct the problem, and a diagram or picture to show why the stall was too small.

Students will probably need 15–20 minutes to compose their letters. Remind them that they should respond with mathematical arguments—that is, they should explain *mathematically* why the stall was too small. Encourage them to be clear and specific about their ideas and suggestions. See the **Teacher Note**, Assessment: The King's Foot (p. 00) for things to look for in student work and for a variety of student responses.

❖ **Tip for the Linguistically Diverse Classroom** Give nonnative speakers the option of writing the letter in their native language, or of concentrating on drawing the diagram or picture that shows what the carpenter could do to correct the problem.

When students have finished their letters, take a few minutes for them to share their ideas with the class; then finish reading the story.

Activity

Introduction to Measuring Tools

In the story we just read, the carpenter solved her problem by making a copy of the king's foot and using that as a measuring tool whenever she built something. Besides using the king's foot, there are other things you can use to measure with. What kinds of tools have you used when you measured, and what kinds of things did you measure ?

Elicit children's ideas about how they have used different measurement tools in different situations; for example:

Why did you use a tape measure, not a ruler, to measure the room?

This emphasizes that certain tools lend themselves to certain situations. Show students different tools (rulers, a tape measure, a yardstick) as they are mentioned.

If you are using rulers marked with inches on one side and centimeters on the other, call attention to both. Explain that students will be using both metric measure and U.S. Standard measure in this unit, but that first they will be measuring with inches and feet.

Similarly, if you are using a yardstick marked on the reverse of a meterstick, explain to students why you have covered the end of the stick beyond 36 inches (as directed on page 11).

Most students probably will not have seen or used an *inchstick* before. Introduce the inchstick as a measuring tool similar to the ruler, in that it is 12 inches long and is marked off in inches, but a tool that is less confusing than many rulers because it doesn't have extra lines. See the **Teacher Note**, Rulers and Inchsticks (p. 48), for further explanation.

Pass out rulers or inchsticks (at least one to each pair). Explain that they are going to collect some data about how big or long a third grader's foot is, and that for tonight's homework, they will be collecting more data by measuring the feet of the people in their homes.

Deciding How to Measure Our Feet Let students decide as a group how they will measure their feet. Explain:

Scientists collect data from measuring things, and sometimes they have to redo a whole experiment because different people measured things in different ways. Let's save ourselves this work and make some decisions first about how we'll measure.

- **Should we measure with our shoes on or off? Would it make any difference?**
- **Where should we put the ruler?**
- **Should we stand up or sit down while we're measuring? Would that make a difference?**
- **What if the length is a little more than an inch-measure? Should we round down, round up, or use fractional parts of an inch?**

Encourage students to try a variety of methods to see how their results for the same foot differ according to the method used. Give them a chance to talk about what they think gives a reasonable measure of foot length. Help them decide how to make the measurement standard for everyone. Write their agreed-on methods on the board for reference.

Ask students to measure their feet again, following the established method, and write their results on a stick-on note or index card. After they have finished measuring, collect their results on the board in an unordered list and save for tomorrow's class:

7 inches	$6\frac{1}{2}$ inches
9 inches	8 inches
7 inches	6 inches
8 inches	8 inches
$9\frac{1}{2}$ inches	

Make sure your list includes data from all students.

Note: After class, make a line plot of this data on an overhead transparency or chart paper. Keep this for use in a Session 2 activity, Comparing Kids' and Adults' Feet (p. 51).

Session 1 Follow-Up

Homework

Students collect more data by measuring the feet of the people in their home. They will need to collect data from at least one adult. Students will need a ruler or inchstick; send these home as necessary. Have them write down the names of the people they measure, how old they are, and how long their foot is.

We have deliberately *not* included a reproducible student sheet for this activity. It is important that students have opportunities to decide how they will organize and record the data they need to collect. As you introduce this homework, spend a few minutes having students discuss and show ways they might record their family foot data.

If you feel that your students need written directions to take home, have them make a list of things to get for each person: name, age, foot length.

Exploring the History of Measurement The history of measurement is fascinating to some students, and opens a real door to the past. Interested students may want to find out more about how our measurement systems have been developed and are enforced. They could look up early definitions of measures: How did the "foot" as we know it evolve, really? How was an acre defined? Many encyclopedias offer information on the evolution of measurements.

Your students may want to write for information about the history of weights and measures in the United States. Address such requests to National Institute of Standards and Technology, Office of Weights and Measures, Room A-617, Gaithersburg, MD 20899; phone (301) 975-2000.

Teacher Note

Assessment: The King's Foot

The students' responses to the question "Why was the stall too small for the new horse?" will help you assess their understanding of the importance of having a standard unit in order to communicate about measurement. As you read through this student work, consider the following:

- Do students use a mathematical argument that makes sense? Do they convey the idea that the carpenter's feet were smaller than the king's feet thus the stall was smaller than the king intended?
- Do students understand that the king and carpenter need to agree on a particular unit of measure?
- Do they use a diagram to help illustrate either the problem or the solution to the problem?

Following are examples of student work that illustrate a variety of responses.

> Dear Carpenter,
>
> When the king measured the beautiful sleek horse he measured the beautiful sleek horse too small because the beautiful sleek horse needed more room because horses need TONS of room.
>
> Love, Tamara

Tamara's letter is an example of a nonmathematical argument. She seemed to be focusing in on the story situation and characters, identifying the king's problem as one of not knowing enough about how much room horses need, thus ordering a stall that was too small.

> Dear Carpenter,
>
> You and the king have different sizes in feet and your feet are smaller so the stall wasn't big enough for the horse. You should have measured the horse to figure out how big the stable should be.
>
> From Dominic

> Dear Carpenter,
> I know why you are in trouble. You are in trouble because your feet are smaller than the king's feet. I know what you can do. You can measure the king's feet then make the stall 9 feet long and 5 feet wide. Do it with string that has numbers on it.
> Love Laurie Jo

Dominic and Laurie Jo are beginning to integrate some of the important components of measuring and sizing standard measures. They recognize that the king's foot and the carpenter's foot are different sizes. They are beginning to recognize the need for a standard measurement in order to successfully communicate.

Continued on next page

Dear Carpenter,
You got in trouble because you had smaller feet than the king and that's why the horse could not fit in the stall. The king measured the stall with his foot and he should have told you how big it was.
P.S. Use the king's feet to measure next time.
From Yvonne

Dear Carpenter,
I think you should have used the king's foot not yours because the king's feet are bigger than yours. Ask for a second chance and admit that your feet are smaller than his.
Then build a new stall using the size of the king's foot.
From Khanh
Please write back if I am right or wrong

Yvonne and Khanh use a mathematical argument and elaborate more fully on the importance of using a standard measure. Both suggest ways of building a new stall by agreeing on a unit to measure with.

You will probably find a range of responses in your classroom. Throughout the unit you will have further opportunities to assess your students' ideas about measurement.

King feet

King stall

Carpenter feet

Carpenter stall

Teacher Note — *Rulers and Inchsticks*

A typical school ruler can be visually confusing to many students because of the profusion of lines. Although the ordinary foot ruler is a good tool when students understand the concept of measurement and see a need for fractional markings, the simpler the better at the beginning. For this reason we recommend the inchstick as an appropriate measuring tool for third graders. Inchsticks are foot-long measuring tools marked in whole inches only.

You can make a set of inchsticks for your class by using the pattern at the end of this unit (p. 111). A 9-by-12-inch piece of tagboard can be cut into nine inchsticks. When laminated, these simple rulers can be used repeatedly. Commercially made inchsticks can be purchased from Dale Seymour Publications.

When inchsticks were used in one third grade class, the students found they could easily tell the lengths of objects by counting inch-squares. They were relieved not to have the "extra" lines found on the school ruler, and the alternating colors of the inch squares made counting even easier. Later, students were able to deal with simple foot rulers. Fractional parts of an inch were introduced gradually, as needed, when the teacher felt that students were ready. Working with whole inches before introducing fractions gave students a firm base for their understanding of measurement. Gradually they came to see that halves and fourths of an inch sometimes add useful information.

We encourage you to keep the measurement tools simple at the start so that they engage all the students in your class. For any given activity, one pair of students may be using inchsticks, another pair the foot ruler. Keep an eye out for the students who can easily handle the more sophisticated tools as well as those who are struggling. The knowledge you gain from these observations will help you to plan follow-up measurement activities.

Kids' Feet and Adults' Feet

What Happens

Students use sketch graphs and line plots to organize the foot data they have collected about the size of their feet and the size of adult feet. They use rulers to measure an object larger than 12 inches and discuss their ways of measuring. Their work focuses on:

- organizing and describing a set of measurement data
- using inches to measure objects bigger and smaller than one foot

Materials

- Rulers or inchsticks (1 per pair)
- Class list of foot lengths (from Session 1)
- Completed Session 1 homework

Ten-Minute Math: Quick Image Dot Patterns Once or twice during the next few days, take a short break outside of math class and do the Quick Images activity at the overhead projector. Choose a dot pattern cut from the Quick Images Dot Pattern transparency. Students will need only a paper and pencil.

Flash the pattern on the overhead projector for 3 seconds.

Students then try to draw the pattern and figure out how many dots they saw.

Flash the pattern for another 3 seconds and let students revise their responses.

Ask how many dots students saw on successive flashes, and how they decided.

For full directions and variations, see pp. 89–90.

Activity

Looking at Foot Length

Call attention to the list of foot lengths students obtained in Session 1.

Yesterday you collected these data about the length of your feet. It's a little hard to tell from our list what the typical foot length is. Work with a partner and find a way to organize these numbers so that you could answer this question: What is the *typical* foot length in our class?

Students work in pairs for about five minutes, making what you might call "sketch graphs" to organize the data showing the distribution of foot lengths in the classroom. For details on the variety of approaches they might take, see the **Teacher Note**, Sketch Graphs: Quick to Make, Easy to Read (p. 53).

When students have finished, ask them to think about what their graphs show.

What can you say about the lengths of your feet? Is there anything unusual about the data? Are there any *outliers*—any one or two numbers that you find surprising?

Looking at our data, what would you say is the typical foot length of our class?

As students talk about what they think is typical, you will hear many different theories and many reasons. Some students will choose the *mode* (the most frequent value) as the typical length. Some will look for a *range* of values. Some will find a "middle" value (the *median*). See the **Dialogue Box**, Discussing Invented Methods for Finding Typical Values (p. 55).

Activity

Describing Adults' Foot Length

Last night you collected more data about foot length by measuring the feet of the people in your home. Let's take a look at your data and see if we can describe what's typical for a grown-up's foot. I'm curious—did anyone find a foot that was exactly a foot long? What about longer than a foot?

Spend a few minutes discussing these questions. Then on the board or overhead, help students organize the data using a line plot. Students may want to share data about other youngsters at home; explain that for now, you would like them to look at just the *adult* feet they measured.

Earlier you organized the data for your own feet by making quick "sketch graphs." Let's organize your data for adult feet on a *line plot*.

Make a list on the board or overhead of the adult foot lengths that the students collected. Students can help you set up the line plot by suggesting the range and checking off each piece of data as you or another student record it.

Once the line plot is complete, have students describe the data much the same way as they described the data about their own feet (in the preceding activity).

As you look at this set of data, what might you say is typical about the length of adults' feet?

You may see some interesting patterns and variation on this line plot; the data represents both adult men and women, and there tends to be more variation in foot size between men and woman than between young boys and girls. It will be interesting to note if your students make this observation.

Comparing Kids' and Adults' Feet Take out the line plot you prepared yesterday for the foot length data for your class, and post it (or put it on the overhead) next to the line plot for adult foot lengths.

I organized the data about your feet on this line plot. Looking at these two line plots together, what comparisons can you make about the length of third graders' feet and the length of adult feet?

As students share, encourage them to support their ideas with reasons based on the data. They might say, for example:

> Grown-ups' feet are longer maybe because they are taller, or because they are older.
>
> Our foot sizes are more spread out (have a bigger range) than grown-up sizes, because our feet are still growing and grown-ups' feet usually stay the same.

Activity

Things Longer Than a Foot

Many confusions about standard measures will surface when students are measuring distances longer than a foot (or a yard). The **Teacher Note**, Using Measuring Tools (p. 54) offers some ideas for handling typical misunderstandings. If any of your students measured someone's foot that was more than 12 inches long, begin this session by asking how they did it. If there were no feet longer than a foot, begin this activity this way:

Often when we measure objects or distances, they are longer than one foot. Measuring them can be a little more complicated than measuring something that is less than one foot. Using your ruler or inchstick, measure the longest side of your desktop.

If your students do not have desks, have them measure something else in the room that is longer than one foot, like a tabletop. They can record their measurement on a piece of paper.

As students measure their desks, circulate around the room and observe their approaches.

- Which end of the ruler do they put where?
- Do they start at zero?
- How do they calculate the total number of inches, given that the desk is longer than the length of the ruler?

- If your students are using rulers that have inches along one edge and centimeters along the other, or on reverse sides, do they mix up or combine the two units?

These are very common mistakes that students will make as they begin to use standard measuring tools.

Ask several students to demonstrate how they measured their desktops. Try to choose students whom you observed using different strategies and approaches. As students demonstrate their measuring techniques, you can highlight important strategies by commenting and questioning as they demonstrate. For example:

I noticed that Latisha made sure to line up the beginning of the ruler with the very edge of the desk.

Cesar used his finger to mark the end of the ruler, and then he moved the beginning of the ruler to that point so he could continue measuring.

I see that Kate is being careful not to leave any space between where the ruler ends and where the next one begins.

Could you explain how you figured that your desk was 21 inches long if you measured that it was 1 foot plus 9 inches?

If your students do not mention the yardstick as a tool that is good for measuring things longer than a foot, you should introduce it as another type of measuring tool. Most likely your students will be familiar with yardsticks, but many will not have used them for real measurement.

At the end of this activity, explain that during the next two sessions, students will be measuring more things in the classroom and collecting data about the things they measure.

Session 2 Follow-Up

More Data Collection Some students may want to conduct more research and collect more data about foot size. Is there another group whose feet they might measure?

Big Feet The *Guinness Book of World Records*, edited by Peter Matthews (New York, Bantam Books, 1993), records foot lengths of the world's tallest known individuals. Robert Perching Wadlow, the world's tallest man at 8 feet 11.1 inches, wore size 37AA shoes that were 18.5 inches long. Zeng Jinlian, the world's tallest woman, had 14-inch-long feet. Some students may want to create a display that shows paper cutouts of feet in these "record" lengths along with cutouts of feet the size of the children and adults they have been measuring.

Teacher Note

Sketch Graphs: Quick to Make, Easy to Read

What exactly is a graph, and what is it used for? Making a graph is often taught as something you do *after* you have analyzed your data. In this sense, graphing is seen as a visual way of communicating what you have found. Certainly, a graph is an effective way to present data to an audience at the end of an investigation. But graphs, tables, diagrams, and charts are also data analysis tools. A real-world user of statistics employs pictures and graphs frequently during the process of analysis as a means of better understanding the data.

Many working graphs need never be shown to anyone else or posted on the wall. Students can make and use them just to help uncover the story of the data. We call such representations used during the process of data analysis "sketch graphs" or "rough draft graphs."

We want students to become comfortable with a variety of such working graphs. Sketch graphs should be easy to make and easy to read; they should not challenge students' patience or fine motor skills. Unlike graphs for presentation, sketch graphs do not require neatness, careful measurement or scaling, use of clear titles or labels, or decorative work.

Sketch graphs:

- can be made rapidly
- reveal aspects of the shape of the data
- are clear, but not necessarily neat
- don't require labels or titles (as long as students are clear about what they are looking at)
- don't require time-consuming attention to color or design

Encourage students to invent different forms until they discover some that work well in organizing their data. Sketch graphs might be made with pencil and paper, with interlocking cubes, or with stick-on notes. Cubes and stick-on notes offer flexibility because they can easily be rearranged.

One standard form of representation that is particularly useful for a first look at the data—the line plot—is suggested for use throughout this unit.

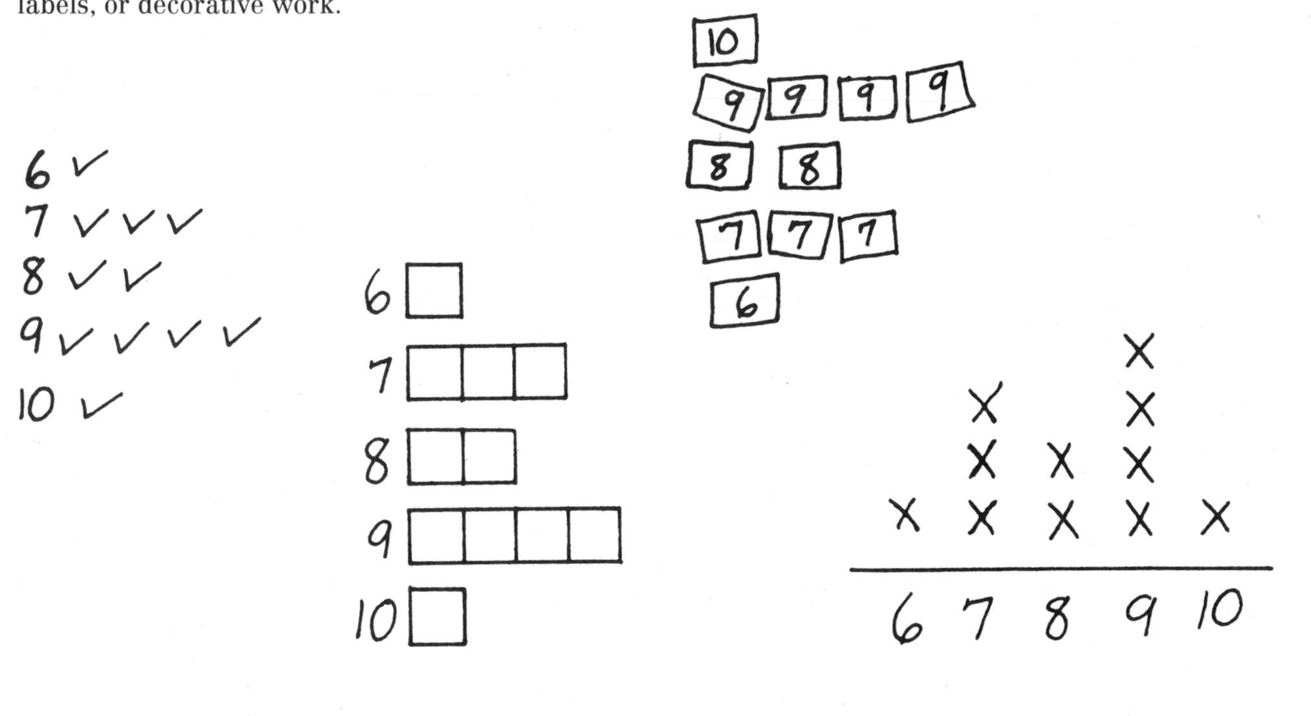

Teacher Note — *Using Measuring Tools*

Measuring seems simple enough, but for elementary students it can pose a real challenge. Even though students can do measurement worksheets and manipulate measurement data on paper, they may not have had much experience using rulers and other measurement tools. Students who have done woodworking, who have built things at home, who have played with and built models (including dollhouses) will be the most expert at this activity. They have some physical experience to draw from—they are familiar with tools and know how to use them, and they may have internalized the sizes of the measurement units.

Conducting measurement activities in small groups allows you to take the observer's role, to see what understandings and misconceptions your students have. Listening to their comments and questions as you circulate will give you some real insight into the mechanical and conceptual problems that measurement often presents to them.

Some predictable problems arise. For example:

- The need to line up the ruler at zero is not always obvious.
- Students may start from the wrong end when they pick up and move a ruler.
- They may combine units, using both metric and U.S. Standard systems.
- They may not notice that their "yardstick" is in fact a meterstick.

All these skills depend to some degree on prior measurement experience.

For many teachers, it helps to think of students' initial measurements as first approximations to their answers, rather than final results. A vital part of their learning is the opportunity to discuss the reasonableness of their measurements, to measure several times, and to correct their measuring mistakes. When students feel the results matter, they become much more precise.

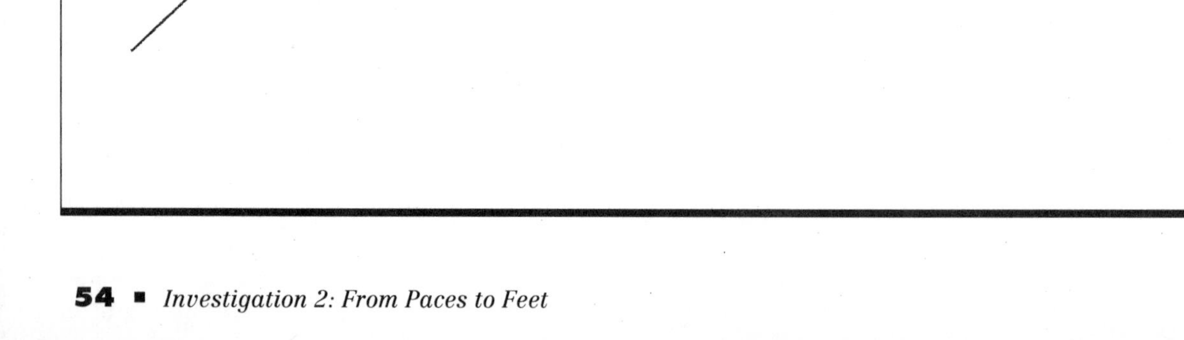

DIALOGUE BOX

Discussing Invented Methods for Finding Typical Values

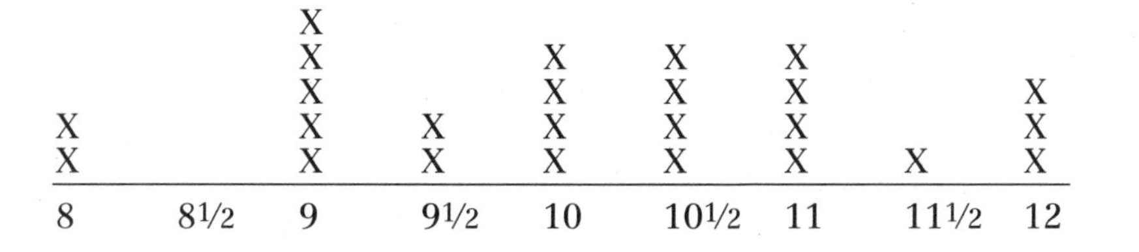

Here the students are considering the line graph they have created with the adult foot-length data they gathered at home. This discussion occurs during the activity Describing Adult's Foot Length (p. 50).

So, looking at these data, how can we find the typical length of an adult's foot?

Rashad: We think you should pick a number that comes up the most, so we got 9 inches, because more grown-ups had 9-inch feet than any other number.

What does everyone else think about that method?

Annie: We did the same thing. There are a lot of 9's, so that seemed like what was typical.

Did anybody make a different choice?

Jennifer: We came out with $10\frac{1}{2}$.

So your choice is a little higher than what Rashad's group picked. Why do you think that's reasonable?

Su-Mei: I don't know. It just seemed like that would be it.

But I'm interested in your reason for $10\frac{1}{2}$. [*Pause. Still no response from the student.*] **I see a big clump of data between 10 and 11 inches.**

Su-Mei: Yeah, the clump seems like it's crowded around $10\frac{1}{2}$.

Yes, that's an interesting reason. I can see your reasons for both of these methods. Does anyone have a good argument for choosing one over the other for the most typical value for the adults' feet?

Sean: Well, even though 9 inches has the most, there are still only five teachers with 9, but there are 12 grown-ups bunched around 10, $10\frac{1}{2}$, and 11.

So you'd choose the biggest clump of data?

Jennifer: Yes. I don't think you should pick the highest or the lowest because that's not like what's typical. Most grown-ups don't have feet that big or that small.

Rashad: Yeah, like with our paces, the typical pace wasn't the shortest or the longest, but somewhere in the middle.

Sessions 3 and 4

Materials

- Rulers, yardsticks, inch-sticks (1 tool for each pair)
- Tape, chalk
- Pattern blocks (5–6 at center)
- Chart paper (3 sheets)
- Student Sheet 5 (1 per student)
- Student Sheet 6 (1 per student)

Measuring Centers

What Happens

In four activities organized in measuring centers around the classroom, students use standard measuring tools to measure familiar objects in the classroom; they locate "benchmarks" on their body, which they can use to estimate lengths without a ruler; and they collect and discuss data about how far they can jump and how far they can blow a pattern block. Their work focuses on:

- using standard measuring tools
- organizing data on a line plot
- describing data

Activity

Working at the Measuring Centers

Four Centers For most of Sessions 3 and 4, pairs of students will be working simultaneously on four measuring activities going on at different centers around the classroom. In all four, students use standard measuring tools in a real context to collect data, which they then organize and analyze.

How to Set Up the Centers The requirements for the four centers are as follows:

Center 1: Measure and Compare—measuring tools, copies of Student Sheet 6 (two pages, stapled)

Center 2: Body Benchmarks—measuring tools, posted chart paper (for students to record their benchmarks)

Center 3: How Far Can a Third Grader Jump?—open space for jumping, a starting line marked with tape, chalk, or paper clips (to mark the length of a jump), measuring tools, a sheet of rules (established by students), posted chart paper (for students to record their data)

Center 4: How Far Can You Blow a Pattern Block?—a table at least 24 inches long, a starting line marked with tape, 5–6 pattern blocks, measuring tools, a sheet of rules (established by students), posted chart paper (for students to record their data)

Centers 1 and 2 can accommodate many students at once, with pairs working at their desks. Center 3 (jumping) can be set up in the hallway or in an open area of the room. Depending on how large an area you have, determine how many pairs of students can work comfortably on this activity at once. You might consider having two areas set up for this activity, thus allowing for more participants at one time; we suggest at least two pairs per area.

Introducing the Measuring Centers Distribute Student Sheet 5, Measuring Center Data Sheet, to each student. Introduce the centers:

Today and tomorrow you will have many opportunities to measure different things in the classroom. Some measurements will be less than a foot, and others will be much longer than a foot. In each situation you can decide which measurement tool is easiest for you to measure with. There are rulers, inchsticks, and yardsticks available. You and your partner will each have a recording sheet [Student Sheet 5] to keep track of your work as you move from center to center. You can visit the measuring centers in any order.

Introduce each activity individually to the whole class. For Centers 3 and 4, you will need to establish rules for measuring; see the following activity descriptions for specifics.

After all the centers have been explained, students choose where they would like to begin. Every student should work at all four measuring centers at some point during these two sessions. Tell students to keep their Data Sheets, because they will be using their data later in the investigation.

Center 1: Measure and Compare

Students measure and compare the size of familiar objects in the classroom, recording their findings on Student Sheet 6.

Some students will be familiar with the terms *length* and *width*, but others will not. Discuss these terms with your students as you introduce this measuring center. The **Dialogue Box**, What's the Length and What's the Width? (p. 62), provides examples of typical student thinking.

Students may find that they need to use fractional parts of an inch when they measure. However, knowing that an object is "about 3½ inches" or "close to 4 inches" is appropriate estimation for third graders.

Center 2: Body Benchmarks

Students find their own personal "benchmarks" on their body that correspond to 1 inch, 6 inches, or 12 inches (1 foot). Once they have established a familiar benchmark, they can always estimate measurements when they don't have a ruler handy. The **Teacher Note**, Benchmarks on the Body (p. 61), discusses this practice further. Post a chart near the center or on the board for students to record their benchmarks.

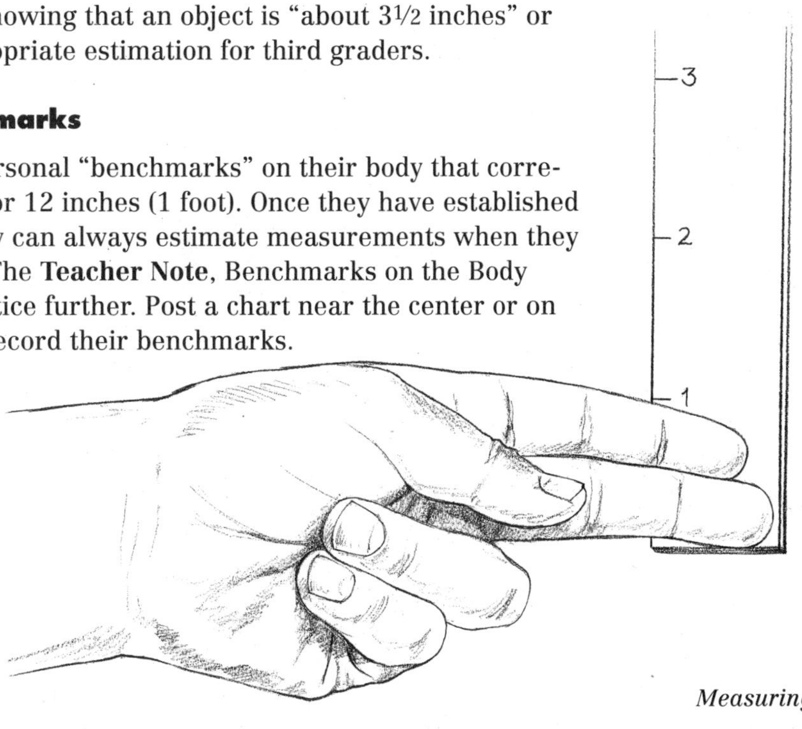

Center 3: How Far Can a Third Grader Jump?

Students measure the length of their jumps from a standing position. In Session 4, they organize these data on a line plot and later discuss it. For this activity, it is important for students to agree upon a method of measuring the distance they can jump. Even with tape on the floor to mark a starting line, you and your students will need to discuss ways of marking where they jumped to and how they should measure that distance. Some students suggest measuring from the starting line to the heel where it lands, while others think the measurement should be to the toe. Most agree on starting with their toes right on the starting line.

The distance jumped can be marked with a piece of chalk or another small object (such as a paper clip or block). Students will need to agree on one way to measure so that all their data represents the same distance. They may also want to discuss how many practice trials they can take before they make their "real" jump. In some classrooms students suggested measuring the best jump out of three tries. Other classes weren't so generous and decided that there were no extra chances, just the real thing. Once the rules are established, write them up and post them at this center for students to refer to.

Also post a piece of chart paper on the wall near the activity for students to record their data (for use at the end of tomorrow's session).

Center 4: How Far Can You Blow a Pattern Block?

This activity gives students another set of measurement data to organize and describe. Students can probably use many of the same rules that they agreed upon for the previous activity, in terms of how and what to measure, and whether to allow any "trial blows." As before, write up the established guidelines and post them for students to refer to. Also post a piece of chart paper nearby for recording their data.

Teacher Checkpoint

Observing as Students Measure

As students are working in the different centers, you will have an opportunity to observe how they use standard tools. Third graders will vary in their skill and use of rulers and yardsticks. Some will be very familiar with these tools and able to measure accurately. Others may know how to use a ruler, laying it end to end, but may be unsure about how to calculate the total distance measured—that is, they may not be sure that 1 foot-ruler-length (12 inches) plus 5 inches more is a distance of 17 inches. Many third graders will have had very little experience using standard measuring tools. For them, understanding where to place a ruler and how to read it will be a challenging starting point.

While students are in the measuring centers, try to observe each pair at work. Consider these matters:

- How does the student orient the ruler?
- What is the student's choice of measuring tools?
- How do students measure *large* distances or objects? That is, do they use one tool that they pick up and move along, or do they use more than one ruler or yardstick to cover the distance?
- How does the student keep track of and calculate the total measurement?

At the end of the first session at the measuring centers, remind students that they will continue with their work during the next math class; thus they should save Student Sheet 5 for recording their work.

Activity

Organizing Our Jumping Data

Begin Session 4 by helping students organize the data that has been collected so far at the "jumping" center. Use an overhead transparency or the chalkboard to set up a line plot.

When we looked at the data about our foot length, we used a *line plot* as a way of organizing the information. Let's do the same to display the data that we are collecting about the length of our jumps. What should we set as the *range* on our line plot?

Take student suggestions about how to set up the line plot. Encourage them to think about the range for the current data and how that might change by the end of today's class. Be sure to leave empty spaces on either end of the line plot so that the range can expand if necessary.

Individual students can transfer their data from the posted chart to the line plot. Students measuring the length of their jump today can then add their data to both the chart and the line plot. The data collected from this activity will be discussed at the end of Session 4.

At the end of today's session, we will talk about your data on How Far Can a Third Grader Jump? Tonight (or tomorrow) you will be organizing the data from How Far Can You Blow a Pattern Block? It's important that we have a full set of data, so if you have not done either of these activities, please choose one of them to begin with today.

As students continue to work at the four measuring centers, you might use a portion of today's session to work with a small group of students who are having difficulty using a ruler successfully.

Activity

Discussion: How Far Can a Third Grader Jump?

About 15 minutes before the end of Session 4, gather the students for a discussion about the jumping data they have collected and recorded on the line plot.

As you look at this line plot, what can you say about how far a third grader can jump? Is there anything unusual or surprising to you about this data?

As students share, record their observations on the board or overhead.

Suppose someone asked you, "What's the typical jump of a third grader?" What would you say?

Encourage students to justify their ideas about the typical jump. As with foot length in Session 2, student's theories and reasons for what's "typical" will vary. Some might focus on a clump of values, and others might choose a single value in the middle of the data. By listening to each other's theories, students will begin to sort out answers that are sensible and for which they have strong ideas. You may want to look back at the **Dialogue Box**, Discussing Invented Methods for Finding Typical Values (p. 55).

What if the third grade across the hall did this activity, and then added their data to our line plot. How do you think the line plot might change?

Benchmarks on the Body

Teacher Note

Carpenters used to make marks on their workbenches to indicate lengths that they often used—such as 2 inches, or 4 inches, or 1 foot. These "benchmarks" speeded up the measuring process, since the carpenter did not then need to find a measuring tape or ruler every time. In a similar way, identifying "benchmarks" on the body is useful because we often need to estimate measurements when we don't have a ruler handy.

You can help your students find such benchmarks. For instance, an inch is about the size of the first joint on the thumb. Ask students to check this out with their rulers or inchsticks to see if it's true for them. They can then use their rulers or inchsticks to find a place on the body that is 1 foot long. One possibility is the forearm. Or, they may find something—like a handspan—that measures 6 inches; two of these would equal 1 foot.

Teachers who have done this—had students find benchmarks on their bodies and then use those benchmarks to estimate other lengths—have found that it helps their students internalize the measures. They are able to estimate measurements with reference to their own bodies and no longer feel the need to rely on a ruler all the time. One third grade teacher overheard his students giving each other directions to pour water into a glass. The first student asked, "How full?" and the second responded, while looking at her thumb, "About two inches."

Once benchmarks on the body are established, it's important to use these measurements as much as possible every day in school. Teachers have found ways to weave measurement into lining up for recess ("Everyone whose little finger is less than 3 inches can line up now"), or doing tasks at the end of the school day ("Find an object that's between 5 and 6 inches long and put it away"). If you have your students first estimate with their benchmarks and later use a ruler to check their estimates, you will be helping them internalize a system of measurement. That skill will help them tremendously, both in mathematics classes and in their daily lives.

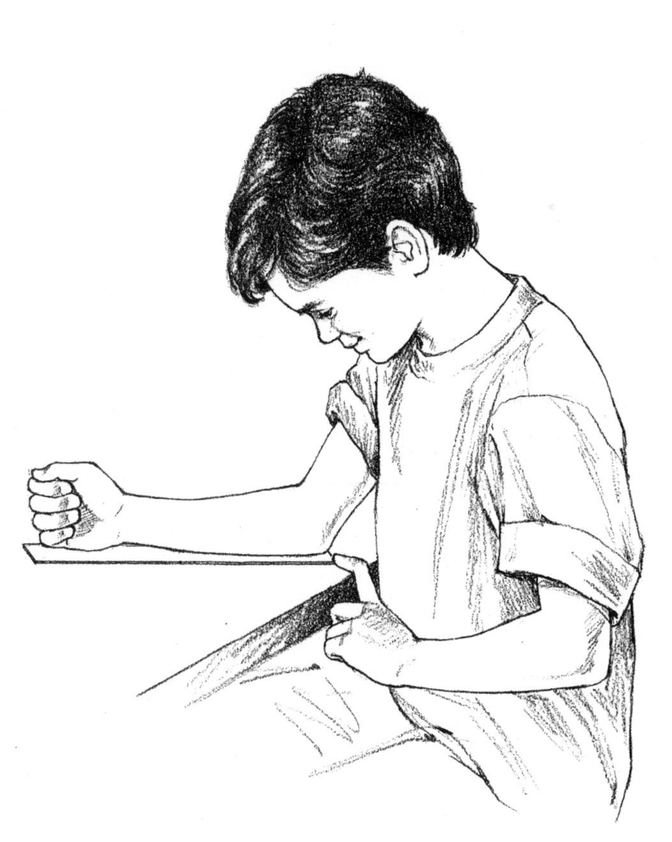

DIALOGUE BOX

What's the Length and What's the Width?

The following discussion took place as the teacher was introducing the Measure and Compare center (p. 57).

One of the comparisons that you'll be doing is comparing the *width* of a dictionary to the *length* of your reading book. I have the dictionary right here. What part should you measure if you want to find the *width*?

Aaron: I think you should measure the part that goes along the bottom.

Tamara: Yeah, because width is how wide something is. It's almost the same word, *width* and *wide*. So I think the bottom edge too.

Elena: I think you should measure this part—the fattest part. [*Elena points to the binding of the book.*]

So, Elena, you're suggesting we measure the distance between the front and back cover?

Elena: Right!

Maria: I don't think that's the width, even though that part is kind of wide. I think that part is called the *thickness* of the dictionary.

Michael: Well, I think that *width* is the widest part of something, so I think you should measure from top to bottom.

Jamal: I disagree with Michael because I think that if you measure from the top to the bottom, that's like measuring how tall it is, and I think you'd be measuring the *length* of the dictionary, not the width. I think the *width* goes from left to right, and the length goes from top to bottom.

Latisha: But what happens to the length and the width when you turn the dictionary the other way so that it's on its side? Then does the length become the width and the width become the length? This is getting confusing.

You're right! I'm feeling a little confused about the length and width, too.

Su-Mei: Hey, I know. Why don't we look it up in the dictionary that you are holding! [*Su-Mei looks up the definition of width and reads it aloud.*] Width—the quality or condition of being wide; the size of something in terms of how wide it is.

Jamal: That doesn't help at all. We already knew that.

The confusion over the definitions of *length* and *width* is not uncommon. Like other mathematical terms, they are somewhat dependent on the context and situation. We tend to think of such terms as having hard and fast definitions when in fact they do not. What's most important is that everyone agree on how to define the length and width so that the terms are used consistently among the group. By allowing students to discuss their ideas, we are helping them to construct their own definitions and theories based on their experiences.

❖ **Tip for the Linguistically Diverse Classroom** In a discussion like this one, encourage students to use pointing and hand motions as they explain their ideas about length and width.

Moving to Metric

What Happens

Students individually analyze the data they have collected at the pattern block measuring center, allowing you to assess their abilities in data analysis. They are introduced to centimeters and meters as a different system of measurement used by most of the world, and they begin to develop centimeter awareness by making their own centimeter and meter measures. They then use these tools on a scavenger hunt in their classroom, looking for things that are about 1 meter and about 1 centimeter long. For homework, they continue the scavenger hunt in their own homes. Their work focuses on:

- learning to describe a set of data that involves measurement
- developing an awareness of centimeters and meters and how big these units of measure are

Ten-Minute Math: Quick Image Geometric Designs Once or twice during the next few days, do the Quick Images activity at the overhead projector. Remember, this activity is not intended to be done during math time.

Choose a design cut from the Quick Image Geometric Design transparency. Students will need only paper and pencil.

Flash the design on the overhead projector for 3 seconds.

Students try to draw the image and figure out how it is put together.

Flash the design for another 3 seconds, and let students revise their drawings.

Finally, reveal the design for final comparisons. Ask students to describe how they saw the image on successive flashes.

For full directions and variations, see pp. 89–90.

Materials

- Students' Data Sheets (from Sessions 3–4)
- World map or globe
- One-centimeter graph paper (1 per student)
- Scissors, tape
- Student Sheet 7 (2 per student)
- Centimeter cubes
- Chart paper (2 sheets)
- Rulers marked with centimeters
- Metersticks

Activity

Assessment

Analyzing the Pattern Block Data

With students looking at their Measuring Center Data Sheets, ask them to read off, in turn, the longest distance they blew the pattern block. Record these data on the board where everyone can see them.

Have students take out a sheet of paper, and ask them to each make a line plot that shows the distances that their class can blow the pattern block.

On the same sheet with the line plot, they should answer these questions:

- Describe your data. What can you say about the distances our class blows pattern blocks?
- What is typical for our class?

Collect these sheets to assess how students are learning to use line plots and analyze measurement data. Use the **Teacher Note**, Line Plot: A Quick Way to Show the Shape of the Data (p. 20) as a guide for assessing students' representations. The **Dialogue Box**, Describing the Shape of the Data (p. 24), will give you a sense of what kinds of work students' might do in their written answers. Ask yourself the following as you review their work:

- Are students attending mainly to individual points in the data (such as their own piece of data), or are they summarizing the data set as a whole? It is difficult for students at this level to attend to and summarize the set of data as a whole, but a goal is to see some movement in this direction.
- Do students describe something about the range, noticing the shortest and longest attempts?
- Do students describe where the clumps and bumps are in the data? Do they notice the holes, or places where there is no data?
- What do students say about the "typical" value? Looking for the middle piece of data is an excellent strategy. You can also expect many students to say that the typical value is the mode, or the most frequently appearing number.

Activity

Introducing Centimeters and Meters

In preparation for this discussion, read the Teacher Note, Background on the Metric System (p. 67). If you have students from other countries, they might be familiar with the metric system and could take a leadership role in these sessions.

Remember in the story, how the king and the carpenter got into trouble with measuring the horse stall? What was their problem?

Review this story briefly, and encourage the students to explain what happens when people don't agree on a standard unit of measure.

Most of you said that there would have been no problem if they had agreed about whose foot to use to do the measuring. People in the United States have agreed to use inches and feet and yards to measure most distances. But other people all around the world have a different system for measuring. It's called the metric system, and it uses units like

centimeters and meters. Only two countries in the whole world use inches, feet, and yards instead of centimeters and meters—the United States and Liberia.

You may want to have students find Liberia on a world map. If they're curious about why these are the only two countries that use the nonmetric (U.S. Standard) system, you might want to give them some background on this (see the Teacher Note, Background on the Metric System, p. 67).

Ask students if they know how big a centimeter is. Anyone who knows can show the class. If no one knows, show the centimeter cubes or the squares on one-centimeter graph paper. Explain that this is the metric unit that we use when we're measuring something small. You might want to hand out the centimeter rulers at this point, and have students find one centimeter on the ruler. If students have been using tools marked with inches on one side and centimeters on the other, they will be somewhat familiar with centimeters already.

Explain that for measuring bigger things, like rooms, we use meters. Show the meterstick. A meter is a little longer than a yard, and it's made up of 100 centimeters—an important "landmark" number.

Activity

A Scavenger Hunt

On one sheet of chart paper, write the heading "Things about 1 cm long"; on the other, write "Things about 1 meter long." Post these where students can write on them.

Distribute a copy of Student Sheet 7, Metric Scavenger Hunt, to each student.

Today you're going to have two scavenger hunts—one at school and one at home. You'll search for things that are about 1 meter long and about 1 centimeter long. In case you don't have the tools you'll need at home, we're going to make a centimeter square and a paper "meter strip" for you to use.

Distribute scissors and one-centimeter graph paper and have everyone cut out a centimeter square. Suggest that they cut the square from the edge of the paper so that they have plenty of paper left to make their meter strip.

Next, ask them how they could make a strip from this paper that is one meter long. Encourage them to use their knowledge of how long and wide the paper is—25 centimeter blocks by 19 centimeter blocks—to figure out how to construct a 100-centimeter strip. (It can be as wide as they want it to be.) This is a good opportunity to review how the number 100 is composed of groups of 20 or groups of 25. These paper measuring tools will be used for the homework. It's better to use actual metersticks and centimeter rulers in class, as the paper cutouts are not very sturdy.

Students spend the rest of the session on their hunt, looking for things in the classroom that are about 1 meter long and 1 centimeter long, recording their findings on Student Sheet 7. The objects don't have to be *exactly* these sizes—encourage students to find things that are *approximately* the right size.

❖ **Tip for the Linguistically Diverse Classroom** Students who are not yet writing in English can sketch the items they find (also for their "home" Scavenger Hunt, as suggested in the Session 5 homework).

When students complete the in-class Scavenger Hunt, they transfer the list from their recording sheets to the two class lists on the chart paper you have posted.

Session 5 Follow-Up

Homework

Give students another copy of Student Sheet 7 and have them take this sheet home, along with the paper centimeter and meter lengths they have cut out. They continue the Scavenger Hunt at home, listing as many objects as they can that are about 1 centimeter or 1 meter long.

Things that are 1 centimeter

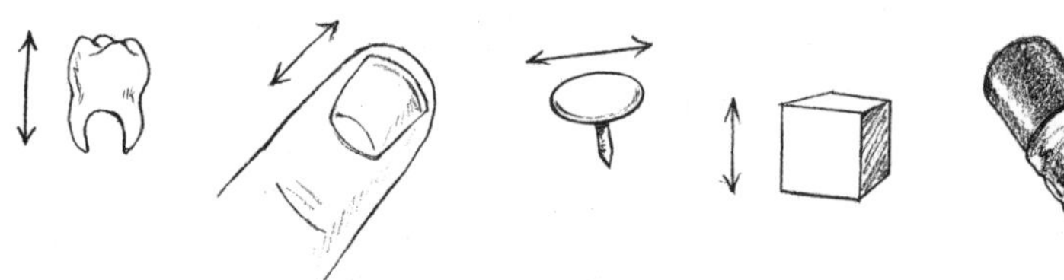

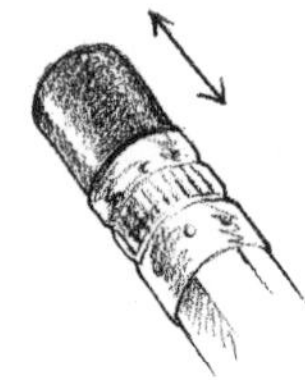

Things that are 1 meter

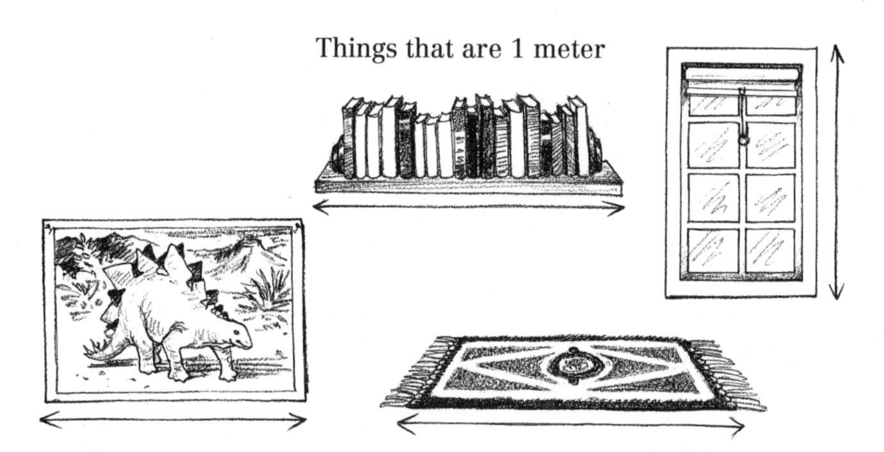

Background on the Metric System

Teacher Note

Only a few countries in the world currently use nonmetric measures of length. Most countries in the world have agreed, at various points in the last century, to use the metric system. Scientists worldwide have agreed to use the metric system so that they can easily compare results and have a common measurement language. Thus scientists in the U.S. are quite familiar with the metric system, but most other citizens are not.

Besides the United States, Liberia makes use of the nonmetric system. Freed slaves from the United States were instrumental in founding the country of Liberia in the early 1800s. One of the customs they brought into their new country was the U.S. Standard system of measurement (sometimes called *customary* or *English measure*). It has remained the commonly used system of measurement to this day.

The United States has resisted many attempts to convert to metric. Even in the 1800s, legislation was introduced to use the metric system. Americans have repeatedly resisted these efforts, in part because the U.S. Standard system is familiar and it is difficult to learn to use a new system. In the 1980s, there was another major effort to introduce the metric system into our lives. Remember when speed limit signs were posted in both kilometers and miles per hour? This new attempt, sponsored by a special Metric Office in the U.S. government, largely failed. Because our country is large and somewhat isolated from those in which the metric system is commonly used, we remain stuck in our ways.

U.S. adherence to the nonmetric system has caused some problems—problems that are both frustrating and amusing. For example, athletes in this country frequently use U.S. Standard measurements in their local, regional, and national competitions. U.S. swimmers typically compete in races that involve multiples of 25 yards, and most pools in this country are 25 yards long, start to finish. But when these swimmers compete in international races, the distances are all multiples of 25 meters. To prepare for international competition, coaches sometimes have their swimmers practice metric races—which of course end somewhere in the middle of the pool!

Sessions 6 and 7

Metric Measurement

Materials

- Metersticks
- Centimeter rulers
- Student Sheet 8 (1 per student)
- Chart paper (7 sheets), markers
- Yarn or string for measuring heads
- Scissors, tape

What Happens

Students conclude their scavenger hunt by examining the data from home and school. Then they make metric measurements related to their clothing sizes. One of these measurements—head size—is used to assess how well students can measure using centimeters. Throughout the sessions, they learn more about meters and centimeters and how to use them. Their work focuses on:

- developing an awareness of centimeters and meters and how big these units of measure are
- practicing measuring skills

Activity

Discussing the Scavenger Hunt

Prepare two large sheets of chart paper, headed "Things that are 1 meter at home" and "Things that are 1 cm at home." As students arrive in class, have them write down the results of their home scavenger hunt on these charts.

Have students finish recording their classroom scavenger hunt results on the posted charts. The home results go on another piece of chart paper. When most students have recorded their lists, briefly discuss what they found.

Was it hard or easy to find things that were about a centimeter long? about a meter long? What are the differences between the things you found at home and at school?

Activity

Body Size Measuring Centers

Set up five stations in the classroom for measuring different sizes. The height center should have at least two places where children can measure, as this is the most time-consuming measurement and tends to go a little slowly. All five centers should be equipped with a couple of metersticks and centimeter rulers. If you don't have enough metersticks and centimeter rulers, you can use centimeter cubes or the paper meter strips the students made in Session 5. The head circumference center also needs yarn or string, scissors, and tape.

Ask if any of the students have ever had clothes that are from another country. If so, ask if they noticed anything about the sizes. Other countries don't use the same sizes we do in the U.S.! If we buy clothes and shoes that are made for use in another country, we often need to know how big we are in centimeters.

If you or a student has any clothing catalogs from other countries, bring them in for discussion.

You're going to find your sizes in centimeters, so that you could buy clothes to fit you if you lived in a different country. Today and tomorrow you will be measuring your sizes for clothes, hats, and shoes in metric. You'll do the work at five measuring centers.

You get to decide what measuring tools to use. There are centimeter measures and metersticks available. You and your partner will each keep track of your work on your own recording sheet. You can visit the measuring centers in any order.

Distribute Student Sheet 8, My Sizes in Metric. Students work in pairs for this activity. Before they start, ask the class to think about how they'll measure different things. It's important that they agree on a common method. There are several decisions to be made for each body measurement. As students agree on certain procedures, write them on chart paper (one sheet for each center) and post for easy reference while students are making their measurements.

- **How will we measure foot size?**
  Ask students if they want to use the same procedures that they used earlier to measure their feet (Session 1, when they were measuring in inches). Remind them what procedures they agreed on (sitting or standing, shoes on or off, and so forth). If they decide on a different method now, everyone must agree.
- **How will we measure our heights?**
  Ask a student to help you demonstrate how to measure height. It could be done lying down or standing up. In either case, encourage them to do it *against* something (floor or wall), as it is quite hard to measure by just holding a measuring tool next to the body.

  This may be the first time students have measured something bigger than a meter, and they are likely to have problems with placement of the meterstick. Two sticks usually work best for beginners, as they can put one on top of the other. Students often aren't concerned about which end they use first, so be sure your demonstration includes some questions about what would happen if you reversed the meterstick:

  Would you get the same measurement? Which would be the correct measurement? Why?

- **How will we measure our hat size (the distance around our heads)?** Explain that hat size is usually measured right around the head, directly over the eyes. But how can they measure around someone's head? Heads are round, and the meterstick is straight. Ask for suggestions. If no one thinks of it, suggest using yarn or string, then measuring the string. Tell students to tape their own "head string" onto their recording sheet. You will use this later to check their measurements for accuracy, in the suggested assessment (p. 71).
- **How will we measure our sleeve length?** This is an important measurement for figuring out shirt size (particularly in long-sleeve dress shirts). It's usually measured from top of the shoulder to the bone in the wrist, with the arm held out straight. Students should discuss how to hold their arms when they make this measurement.

  Does it make a difference if you bend your arm or hold it straight?
- **How will we measure our pants length?** For pants length, measure from the waist down to the ankle bone. Students should decide whether they'll measure from the side, front, or back. Again, they should discuss how they will stand (or sit) when this measurement is being taken.

Once you have introduced the five measurements and written up the procedures, students begin working in pairs to collect the data.

Activity

Discussing Our Height Findings

After most students have completed the body measurement activities, call them together to discuss their findings on height. Using either an overhead transparency, the chalkboard, or chart paper, record each height (in centimeters) as the students read aloud from their recording sheets. Then ask the class to help you make a line plot of these data.

What should I use for my lowest number on the line plot? How about the highest number? Should I mark each number in between?

Some children will probably suggest that you only write down the numbers that are in the data set—for example, 125, 128, 130, 131, 132, 134—and skip the numbers in between. This is a good point to discuss. Ask them whether this system would work if any absent children in the class came back and recorded their heights, or whether it would work if another third grade class added their data. This is a good opportunity to talk about "holes" in the data, and how these should be represented.

Students will probably also notice that their height data are more spread out in centimeters than they are in inches. Many children like this precision—Kate may be about the same height in inches as Liliana, but may discover with delight that she is actually one centimeter taller!

Another thing students may talk about is how they made the "second part" of the height measurement. That is, once they got 100 centimeters, they put the meterstick up again, and got a second number. Ask what they did with the two numbers (for example, 100 and 29). Was it easy to put them together? If they don't discover it on their own, point out that one real advantage of the metric system is that it's easier to put measurements together, because we're starting from 100 or multiples of 100.

Assessment

Checking Students' Metric Sizes

Student Sheet 8, My Sizes in Metric, offers a good opportunity for you to check to see how well students are measuring. While you will not necessarily know if their measurements are all accurate, check to see that they are in the ballpark. Most third graders' measurements will be in these ranges:

- Height: 125 to 140 cm
- Foot length: 16 to 23 cm
- Sleeve length: 40 to 48 cm
- Pants length: 62 to 70 cm
- Head diameter: 50 to 55 cm

If any students' responses seem out of range, check with them individually to see how they did their measurements.

It's fairly easy to check the accuracy of their head circumference data by simply measuring each piece of string (attached to the student sheets) yourself to see if it corresponds to the length that the student recorded. If there is a discrepancy, check with the student to figure out why. Observe as the student shows you how he or she measured.

INVESTIGATION 3

Measuring Project: Do Our Chairs Fit Us?

A Note on Investigations 3 and 4

Investigations 3 and 4 offer two final projects for this unit. You may do both, or you might select the one that best fits your situation and the interests of your students.

The first project, Do Our Chairs Fit Us? can be done using either metric or U.S. Standard measure. Students focus on their classroom furniture, using measurement to analyze whether the classroom furniture fits them and making recommendations about the optimal distribution of chairs for the class. If your students show an interest in measurement as a tool to analyze their own sizes (such interest will have emerged in the foot-size and height investigations), this investigation will be very appealing to them.

In the second project, Balobbyland, students work in pairs to design living environments for a group of mythical creatures called Balobbies (pronounced Bah-LOB-bees.) They use one-centimeter graph paper to construct different spaces for the Balobby village. Experience with counting and using centimeters is stressed throughout, along with constructing spaces of appropriate sizes. This investigation encourages students to stretch their imaginations and deepen their experience with metric measure.

What Happens

Session 1: What's a Good Fit? In this session students first decide what it means to have a chair that fits. They establish a connection between chair height and leg length. They then collect data by measuring the height of their chairs and the length of their legs.

Session 2 and 3: Do Our Chairs Fit Us? Students work with a partner to organize, represent, and analyze the class data collected in Session 1. They use this information to determine whether the chairs in their classroom are a good fit, and they make a recommendation to the principal about their findings.

Mathematical Emphasis

- Using standard measures (either metric or U.S. Standard) in more complex situations in order to gather and analyze data concerning size and proportion

What to Plan Ahead of Time

Materials

- Rulers, inchsticks, and yardsticks; or metersticks—depending on the measurement system you choose (all sessions)
- Materials for presentation graphs and written work: paper, pencils, graph paper of various sizes, markers, stick-on notes, colored dot stickers, scissors, and construction paper (Sessions 2–3).

Other Preparation

- Duplicate a class list for each small group to use for recording their data (Session 1).
- Borrow a school furniture catalog.
- Borrow a desk chair with an adjustable seat (the kind that moves up and down).
- Collect chairs in three sizes: very small (kindergarten size), the size your students use, and adult size.

Session 1

What's a Good Fit?

Materials

- Rulers, inchsticks, and yardsticks; or meter-sticks
- Class list (1 per group)
- Chairs in three sizes
- Chair with an adjustable seat
- School furniture catalog

What Happens

In this session students first decide what it means to have a chair that fits. They establish a connection between chair height and leg length. They then collect data by measuring the height of their chairs and the length of their legs. Their work focuses on:

- using measurement as a way of collecting data
- conducting a data analysis project

Activity

Looking at How Chairs Fit

Students' school desks and chairs are often ill-fitting. Use of furniture tends to evolve over time in a classroom, and students accept what's there without much concern over its fit. Adults' office furniture is taken more seriously. Fitting chairs and desks to people is part of an important field of work, called *ergonomics*. Once students realize the possibilities, they are generally interested in improving the fit of their working furniture.

Today we're going to start work on a real-life measurement problem. Fitting chairs and desks to people has become an important field of work. People who design chairs think a lot about the size of the chair and the size of the people who will use it.

Two of the most important measurements that they pay attention to are the height of the chair and the length of a person's leg. Why do you think these two measurements are important to how a chair fits?

To illustrate this, set three chairs before the class—one that's much too small, one that's about right, and one that's too big. Ask a very small student to sit in the biggest one. Explain that when we are sitting in a chair, a good fit is determined by sitting with our back against the chair back.

Does this chair fit Saloni? How can you tell by looking? What do you notice about Saloni's legs compared to the height of the chair? What part of the leg is most important when it comes to fitting a chair?

Ask the same child to sit in each of the other chairs. Have students discuss the "fit" of each chair, answering the same questions. Help students focus on the length of the leg from the knee to the ground if they are not doing so already. For a typical discussion, see the **Dialogue Box**, Does This Chair Fit? (p. 79).

Ask a few other children of different sizes to try each chair. You might continue to dramatize the matter of "fit" in a chair, focusing on chair height and leg length, until you think your students have some ideas about how the two are connected.

If all the chairs in your classroom are uniform in size, then have students focus only on how people of different sizes fit in those chairs.

Has anyone ever seen a chair with a seat that can be raised or lowered? Why do you think a chair would be designed this way?

If you were able to borrow a chair that can be adjusted, show it to the students.

When some schools buy chairs for classrooms, they buy chairs of different heights. Why do you think they do that? Other times, a school might buy chairs that are all the same height. Which do you think is a better idea? Why?

Schools frequently get furniture catalogs that offer chairs in different styles and sizes. If you can borrow a such a catalog, share it with your students.

Activity

Measuring Ourselves and Our Chairs

During the next three math classes, we are going to do an investigation about the chairs in our classroom and whether or not they fit us. You'll work with a partner to collect data about the height of your chairs and the length of your legs from your knee to the floor. Once all the data for our class are collected, you and your partner will work together and figure out a way to organize this information so that you can determine whether or not our chairs are a good fit.

Make available rulers, inchsticks, and yardsticks if you have decided to use U.S. Standard measures, or metersticks if you will be using metric measures.

Students need to collect two pieces of numerical data: the height of their chairs and the length of their leg from the knee to the floor. Each pair will measure their own two chairs and their legs. Spend time discussing with students how they will take these measurements. Remind them that whenever we are collecting data, it's important that everyone do it in the same way so that the information is accurate and meaningful.

Student pairs take their measurements and record their data on a piece of paper. While students are collecting their measurement data, circulate and observe how they use the tools and record their information.

At the end of this session, collect the data and explain that you will put all the information on a class list; then tomorrow each team will have a complete set of data. Don't forget to compile this data and make a copy for each pair before Session 2.

If there is time at the end of this session, students can begin to brainstorm how they might organize their class data to make a determination about the "fit" of the chairs in their classroom.

Sessions 2 and 3

Do Our Chairs Fit Us?

What Happens

Students work with a partner to organize, represent, and analyze the class data collected in Session 1. They use this information to determine whether the chairs in their classroom are a good fit, and they make a recommendation to the principal about their findings. Their work focuses on:

- organizing, representing, and analyzing data
- making recommendations based on conclusions drawn from data

Materials

- Copy of class data (1 per pair)
- Paper, pencils, graph paper, markers, stick-on notes, colored dot stickers, scissors, construction paper

Activity

Making a Clear Picture of the Data

Distribute the class data to each student pair. They will need access to graphing materials such as markers, paper, stick-on notes, colored dot stickers, and whatever other supplies you have.

Today and tomorrow, you and your partner are going to work on figuring out a way to organize and represent the data that we have collected. Your goal is to get a clear picture of the data, so you can decide if the chairs in our classroom fit us.

You will then use this information to make recommendations to the principal about the chairs in our classroom. You and your partner will work together on making a representation of the data and writing a letter to the principal.

These two sessions can be treated as workshops—a time when student teams are working independently at their representations, discussing their findings, and deciding what recommendations they might make to the principal about the classroom chairs.

❖ **Tip for the Linguistically Diverse Classroom** Students with limited English proficiency could be paired with students proficient in English to complete this task.

Begin with a short discussion about how student might approach this work. Questions like the following might help them focus on the task:

How will you organize the data?
What size chairs do we have?
What size legs do we have?

Some students will need more help than others at getting started.

Activity

Making Our Report

The production of a final report (letter to the principal) and accompanying graphs will take the last session of this project. The important point is that the students draw some conclusions from their work and make some recommendations based on the data they have collected.

❖ **Tip for the Linguistically Diverse Classroom** Students with limited English proficiency can provide the visuals (the data, an illustration of their recommendation) to accompany the letter.

You can help students reach their conclusions with questions like these:

Do you feel enough of our chairs fit us well?

What would the principal need to know about what you've been doing? How can you describe it?

What are some realistic recommendations that you could make based on your findings?

In classrooms that have done this project, some students have recommended that they try to switch chairs that are too big or too small with a younger or older class.

Other ideas that might evolve from this investigation are arranging for students to meet with the person who orders furniture for the school, so they can make recommendations for future chair orders. If students conclude that, in fact, the chair situation in their classroom is adequate, they could report the good news that their chairs fit them quite well!

DIALOGUE BOX

Does This Chair Fit?

In this discussion, early in the investigation, the teacher is working hard to help students identify what parts of the chair and body could be compared and measured in order to determine fit, without actually telling them.

We've found that Maria does real well in this chair. Why is that? What matters when we look at making the chair fit, or you fit in the chair?

Sean: The body.

What do you mean about the body?

Maya: She's small.

She's small. But I could say Ricardo's small, too, but he doesn't fit well in that chair. [*Ricardo sits in Maria's chair. He's too big. Everybody giggles.*]

Laurie Jo: He doesn't fit.

But why?

Yoshi: He's bigger.

But what does that mean, he's bigger?

Mark: Because his legs and his knees are sticking up. His legs are too long. I think the middle chair would fit him better. [*Ricardo moves to the middle chair.*]

Dylan: Now it's just right. See, his feet are on the floor.

What else is just right about the way Ricardo is sitting in that chair?

Ly Dinh: Look at his legs—they are straight, and his knees are even with the seat of the chair.

[*Later: Maria and Ricardo are both sitting in chairs that are too big for them, and the teacher is in a chair that is too small.*]

Annie: Maria's feet aren't touching the floor, and Ricardo's feet aren't, and yours are.

So, if her feet are off the floor, does it mean the chair is too big or too small?

Yvonne: Too big.

INVESTIGATION 4

Measuring Project: Balobbyland

What Happens

Sessions 1, 2, and 3: Making a Small World Students work in pairs to design school and home spaces for the Balobbies, a group of mythical and very tiny people who live in the centimeter-oriented Balobbyland. Students use centimeter graph paper to design and construct different spaces for the Balobbies. They label the sizes of these spaces, write about them, and put them together in a Balobby village. Experience with counting and using centimeters is stressed throughout, along with constructing spaces of the appropriate size.

Mathematical Emphasis

- Developing an awareness of centimeters and how to use centimeters to measure
- Using centimeters in a more complex situation—constructing a fantasy land, where the relative size of objects is important

What to Plan Ahead of Time

Materials

- Toy figures, 5–8 cm tall (students might bring these from home). Alternatively, have clay available for making Balobbies.
- Teddy bear counters (optional)
- Centimeter rulers (1 per pair); substitute centimeter blocks, paper centimeter squares, or paper meter strips as needed
- Scissors
- Markers or crayons
- Tape or glue
- Tagboard, 18 by 24 inches: (1 per pair)

Other Preparation

- Collect the plastic figures from students or have them make Balobbies from clay, 1 per student. All should be similar in size, 5–8 cm tall. Some classes have used people from Playmobil® sets, about 7 cm, and from LEGO® sets, about 5 cm. You may want to read the description of Balobbyland (p. 83) ahead of time, so students know why they are bringing in or making the figures.
- Duplicate student sheets and teaching resources, located at the end of this unit, as follows:

  Student Sheet 9: Balobby Plans: 3–4 per pair

  Balobby Space Cards: (pp. 107–110) 1 set per pair, cut apart

  One-centimeter graph paper (p. 112), copied on white paper, 3–4 per pair; also on colored paper, 3–4 per pair

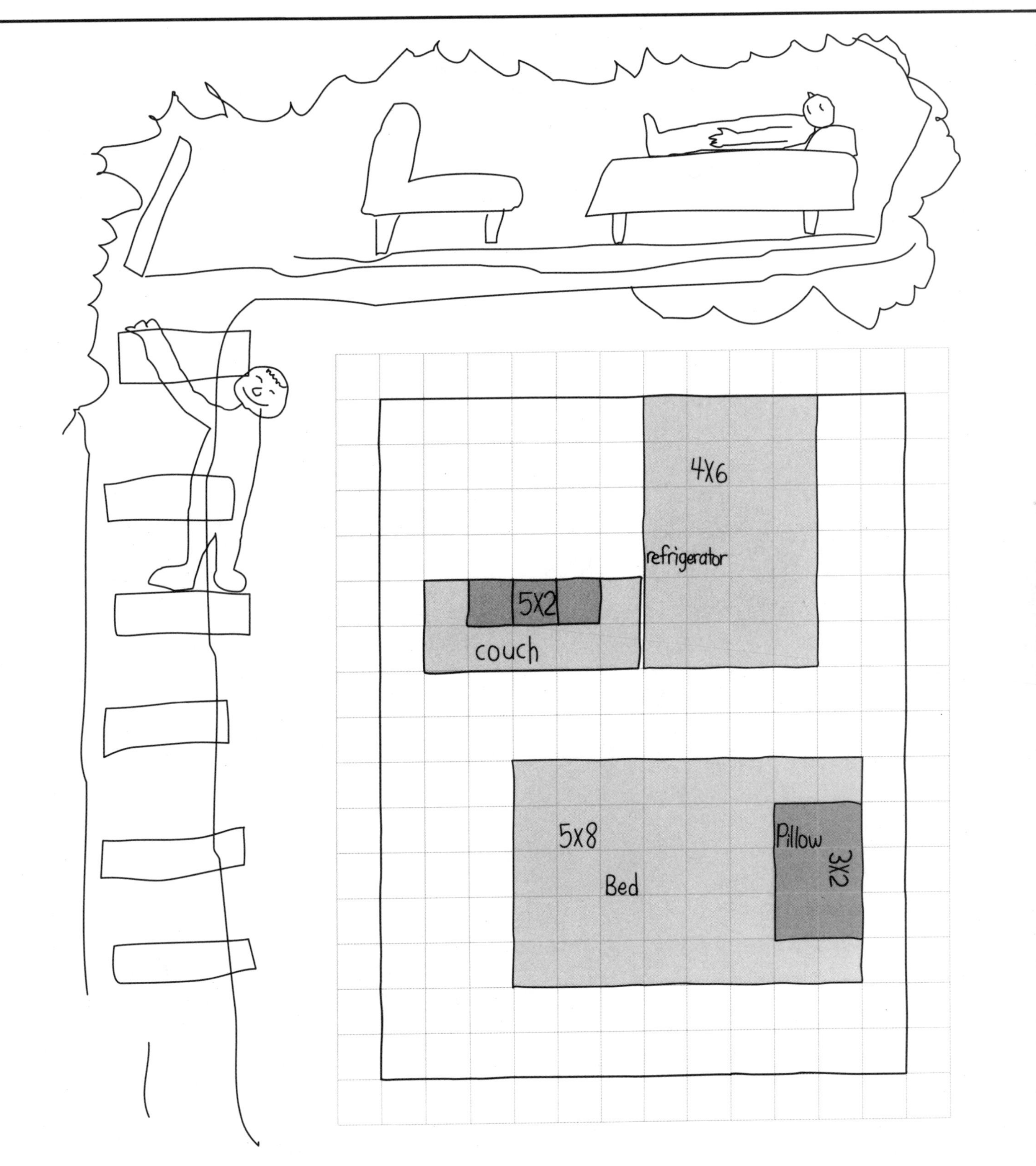
4X6
refrigerator
5X2
couch
5x8
Bed
Pillow
3X2

Sessions 1, 2, and 3

Making a Small World

Materials

- Balobby figures
- Teddy bear counters (optional)
- Centimeter blocks or paper centimeter squares
- Centimeter rulers (1 per pair)
- Scissors
- Markers or crayons
- Tape or glue
- Tagboard
- Student Sheet 9 (3–4 per pair)
- Balobby Space Cards (1 set per pair, cut apart)
- One-centimeter graph paper (white and colored)

What Happens

Students work in pairs to design school and home spaces for the Balobbies, a group of mythical and very tiny people who live in the centimeter-oriented Balobbyland. Students use centimeter graph paper to design and construct different spaces for the Balobbies. They label the sizes of these spaces, write about them, and put them together in a Balobby village. Experience with counting and using centimeters is stressed throughout, along with constructing spaces of the appropriate size. Their work focuses on:

- measuring with centimeters
- using measurement to construct areas of different sizes

Activity

Meeting the Balobbies

Have students get out the figures they have brought in or made; each student should have one. Distribute one copy of Student Sheet 9, Balobby Plans, to each pair. Also make available the centimeter blocks or rulers. Introduce the project:

The Balobbies [*bah-LOB-bees*] **are little beings who live in make-believe Balobbyland. Of course, because they don't live in either the U.S. or in Liberia, the Balobbies use the metric system. They are rather small** [*show one of them*], **so they measure their rooms and furniture in centimeters.**

Over the next few days, you are going to create some places for the Balobbies to live and go to school. The room plans you create will be like architect's plans, on flat paper. You'll need to write down the right sizes so that the architect will know how big to make each space or piece of furniture.

❖ **Tip for the Linguistically Diverse Classroom** Use visual aids to enhance comprehension of the project situation: a world map; a centimeter ruler; a picture of a house interior, with furniture, or dollhouse furniture; architect's plans (or floor plans in home decorating magazines). Lay out a sheet of graph paper and walk a Balobby figure around on it. Then use the ruler to measure out a section of the paper, as well as any toy furniture, to give students a general idea of the upcoming task.

Ask students how tall they think their own Balobby is in centimeters. Sizes will vary, and the next step is for students to measure their Balobbies.

Use rulers or the centimeter blocks to measure how tall your Balobby is. When you're done, write down its height on your Balobby Plans (Student Sheet 9).

Activity

Making Tree House Platforms

The students work in pairs. Each pair needs 2–3 sheets of graph paper, scissors, tape, markers, and the Balobby Plans sheet. To start, everyone will make the first Balobby space—a platform for a tree house. In the next activity, pairs will choose the spaces they want to work on.

A favorite place for Balobbies to hang out is in tree houses. Their tree houses sit on platforms that are about 12 centimeters by 15 centimeters. [*Write 12 cm × 15 cm on the board for children to see the notation.*]
12 cm × 15 cm means that one side is 12 centimeters long and the other side is 15 centimeters long.

Have each pair make a tree house platform of the correct size, using the centimeter graph paper. (They could even cut out a hole in the middle for the tree.) On their Balobby Plans sheet, they should write down what space they're making (a tree house) and how big it is.

❖ **Tip for the Linguistically Diverse Classroom** Sketch a simple tree house, or ask for volunteers to draw their ideas for a tree house on the board, to help clarify the task. Emphasize that their drawings will show a floor plan for the tree house platform. Students may draw rather than write about what the Balobbies do in this space.

At first some students will have trouble seeing the platform from a "bird's eye" perspective. Encourage them to think about how things look from above, and remind them that things need to be drawn flat on their paper. As a visual example, you might sketch the general floor plan of your classroom on the board, drawing key items (desks, chairs, tables, bookshelves) from the overhead "flat" perspective—as squares and rectangles.

When students have finished their plans for the tree house platform, show one or two to the class. Students then furnish their tree houses as they please. You might ask them to brainstorm what furniture the Balobbies would want in their tree houses. Remind students that these things must fit in the tree house space and should be useful to people of Balobby size. Each item should be drawn as a rectangle. Write one or two suggestions on the board, then ask how big these things should be, in centimeters. Next to each item, write the dimensions that the students suggest. For example:

trap door   2 cm × 3 cm

trunk to keep rope   4 cm × 2 cm

Give each pair a chance to figure out one or two pieces of furniture to put in their own tree house. They also need to decide the dimensions for each piece. On their Balobby Plans sheet, they list each piece of furniture and its dimensions. Then they can write down a couple of sentences about what happens in this room—what the Balobbies do in this space. The furniture should be cut from the colored graph paper, labeled, and glued or taped onto the tree house platform, wherever the students want it to go.

Activity

Making More Balobby Spaces

Distribute the sets of Balobby Space Cards, one to each pair.

Over the next couple of days, you will be working to make different spaces where Balobbies can live and play. Then you'll put some of these spaces together on a big piece of tagboard to make a neighborhood.

Students work in pairs to select, create, and write about several spaces for the Balobbies. The Balobby Space Cards will give them some good ideas; encourage them to design their own spaces as well. They could write descriptions for more rooms in the house or school, or for different outdoor spaces. Each pair should plan on making at least three different spaces.

Note: The card for the Teddy Kennel explains that Balobbies keep teddy bears for pets. If you have teddy bear counters, they are great props for students planning this space.

❖ **Tip for the Linguistically Diverse Classroom** Before students begin work, you may want to read through the Balobby Space Cards with the class, using visual aids and quick sketches to help with comprehension of unknown words. Students could add simple sketches to the cards as reminders of the content. In describing the spaces they create, partners with limited English proficiency should be encouraged to communicate their ideas through pantomime actions, drawings, and whatever communication is possible. Partners who are proficient in English can do the writing component of the task.

During these sessions, make available extra copies of Student Sheet 9 as students fill out complete Balobby Plans for each space they make. This requires measuring each object to go in the space, recording its dimensions, and describing how the objects are used in the space (in answer to "What happens in this room?").

The teacher's role during these sessions is to circulate among students, asking questions about the space being constructed and the sizes of different objects in the space.

Activity

Planning a Neighborhood

During the final session, distribute one large piece of tagboard to each pair. Explain that students will be putting some of their spaces together to construct a bigger area for the Balobbies.

You all have some interesting spaces where the Balobbies can live, play, or go to school. Now, we're going to put the spaces together. Think of making a neighborhood where the Balobbies live.

Arrange your places on the tagboard in a way that makes sense. Then draw in sidewalks or roads between any of the places you want to connect. Be sure to label sidewalks or roads with their lengths. The Balobbies will want to know how far it is from one place to another.

Some pairs of students may want to join with another pair in making even larger neighborhoods. Remind students to connect the places with roads or sidewalks, and to measure and label each distance. Some students will prefer to do their measuring with centimeter cubes, while others will use a ruler or a string (especially useful for measuring curved paths). This is another good opportunity to observe their measuring skills.

When the neighborhoods are finished, post the final products and give students a chance to examine each other's work.

Activity

Choosing Student Work To Save

As the unit ends, you may want to use one of the following options for creating a record of students' work on this unit.

- Students look back through their folders or notebooks and write about what they learned in this unit, what they remember most, what was hard or easy for them. You might have students do this work during their writing time.
- Students select one or two pieces of their best work. You also choose one or two pieces of their work to be saved in a portfolio for the year. You might include the student assessment, The King's Foot (Investigation 2, Session 1), and any other assessment tasks from this unit. Students can create a separate page with brief comments describing each piece of work.
- You may want to send a selection of work home for parents to see. Students write a cover letter, describing their work in this unit. This work should be returned if you are keeping a year-long portfolio of mathematics work for each student.

Estimation and Number Sense

Basic Activity

Students mentally estimate the answer to an arithmetic problem that they see displayed for about a minute. They discuss their estimates. Then they find a precise solution to the problem by using mental computation strategies.

Estimation and Number Sense provides opportunities for students to develop strategies for mental computation and for judging the reasonableness of the results of a computation done on paper or with a calculator. Students focus on:

- looking at a problem as a whole
- reordering or combining numbers within a problem for easier computation
- looking at the largest part of each number first (looking at hundreds before tens, thousands before hundreds, and so forth)

Materials

Calculators (for variation)

Procedure

Step 1. Present a problem on the chalkboard or overhead. For example:

9 + 25 + 11

Step 2. Allow about a minute for students to think about the problem. In this time, students come up with the best estimate they can for the solution. This solution might be—but does not have to be—an exact answer. Students do not write anything down or use the calculator during this time.

Step 3. Cover the problem and ask students to discuss what they know. Ask questions like these: "What did you notice about the numbers in this problem? Did you estimate an answer? How did you make your estimate?"

Encourage all kinds of estimation statements and strategies. Some will be more general; others may be quite precise:

"It's at least 35 because I saw 25 and a number in the tens."

"I think it's less than 100 because 25 was the biggest number and there were only three numbers."

"I think it's 25 + 20 because I saw the 9 + 11 and that's 20 and then add on 25 and that gets you to 45."

Be sure that you continue to encourage a variety of observations, especially the "more than, less than" statements, even if some students have solved it exactly.

Step 4. Uncover the problem and continue the discussion. Ask further: "What do you notice now? What do you think about your estimates? Do you want to change them? What are some mental strategies you can use to solve the problem exactly?"

Variations

Problems That Can Be Reordered Give problems like the following examples, in which grouping the numbers in particular ways can help solve the problem easily:

6 + 2 − 4 + 1 − 5 + 4 + 5 − 2

36 + 22 + 4 + 8

112 − 30 + 60 − 2

654 − 12 + 300 + 112

Encourage students to look at the problem as a whole before they start to solve it. Rather than using each number and operation in sequence, they see what numbers are easy to put together to give answers to part of the problem. Then they combine their partial results to solve the whole problem.

Problems with Large Numbers Present problems that require students to "think from left to right" and to round numbers to "nice numbers" in order to come up with a good estimate. For example:

Continued on next page

130 + 243 + 492

$$\begin{array}{r} \$5.13 \\ \$6.50 \\ +\ \$3.30 \\ \hline \end{array}$$

$3.15 × 5

8 + 664 + 130

Present problems in both horizontal and vertical formats. If the vertical format triggers a rote procedure of starting from the right and "carrying," encourage students to look at the numbers as a whole, and to think about the largest parts of the numbers first. Thus, for the problem 130 + 243 + 492, they might think first, "About how much is 492?—500." Then, thinking in terms of the largest part of the numbers first (hundreds), they might reason: "200 and 500 is 700, and 100 more is 800, and then there's some extra, so I think it's a little over 800."

Is It Bigger or Smaller? Use any of the kinds of problems suggested above, but pose a question about the result to help students focus their estimation: "Is this bigger than 20? Is it smaller than $10.00? If I have $20.00, do I have enough to buy these four things?"

Using the Calculator The calculator can be used to check results. Emphasize that it is easy to make mistakes on a calculator, and that many people who use calculators all the time often make mistakes. Sometimes you punch in the wrong key or the wrong operation. Sometimes you leave out a number by accident, or a key sticks on the calculator and doesn't register. However, people who are good at using the calculator always make a mental estimate so they can tell whether their result is reasonable.

Pose some problems like this one:

> I was adding 212, 357, and 436 on my calculator. The answer I got was 615. Was that a reasonable answer? Why do you think so?

Include problems in which the result is reasonable and problems in which it is not. When the answer is unreasonable, some students might be interested in figuring out what happened. For example, in the above case, I accidentally punched in 46 instead of 436.

Related Homework Options

Problems with Many Numbers Give one problem with many numbers that must be added and subtracted. Students show how they can reorder the numbers in the problem to make it easier to solve. They solve the problem using two different methods to double-check their solution. One way might be using the calculator. Here is an example of such a problem:

$$30 - 6 + 92 - 20 + 56 + 70 + 8$$

Quick Images

Basic Activity

Students are briefly shown a picture of a geometric design or pattern, then draw it by developing and inspecting a mental image of it.

For each type of problem—2-D designs or dot patterns—students must find meaningful ways to see and develop a mental image of the figure. They might see it as a whole ("it looks like a four pointed star"), or decompose it into memorable parts ("it looks like four triangles, right side up, then upside down, the right side up, then upside down"), or use their knowledge of number relationships to remember a pattern ("there were 4 groups of 5 dots, so it's 20").Their work focuses on:

- organizing and analyzing visual images
- developing concepts and language needed to reflect on and communicate about spatial relationships
- using geometric vocabulary to describe shapes and patterns
- using number relationships to describe patterns

Materials

- Overhead projector
- Overhead transparencies of the geometric figures you will use as images for the session; we have provided two transparency masters to get you started. To use the images on the masters, first make a transparency, then cut out the separate figures and keep them in an envelope. Include the numbers beside the figures because they will help you properly orient the figures on the overhead.
- Pencil and paper

Procedure

Step 1. Flash an image for 3 seconds. Show a picture of a geometric drawing or a dot pattern. (See below for specific suggestions related to these two options.)

It's important to keep the picture up for as close to 3 seconds as possible. If you show the picture too long, students will build from the picture rather than their image of it; if you show it too briefly, they will not have time to form a mental image. Suggest to students that they study the figure carefully while it is visible, then try to build or draw it from their mental image.

Step 2. Students draw what they saw.
Give students a few minutes with their pencil and paper to try to draw a figure based on the mental image they have formed. After you see that most students' activity has stopped, go on to step 3.

Step 3. Flash the image again, for revision.
After showing the image for another 3 seconds, students revise their drawing, based on this second view.

It is essential to provide enough time between the first and second flashes for most students to complete their attempts at drawing. While they may not have completed their figure, they should have done all they can until they see the picture on the screen again.

When student activity subsides again, show the picture a third time. This time leave it visible, so that all students can complete or revise their solutions.

Step 4. Students describe how they saw the drawing as they looked at it on successive "flashes."

Variations

In this unit you will find transparency masters for two types of Quick Images: geometric designs and dot patterns. You can supplement any of these with your own examples or make up other types.

Continued on next page

Quick Image Geometric Designs Use the Quick Image Geometric Designs transparency. When students talk about what they saw in successive flashes, many students will say things like "I saw four triangles in a row." You might suggest this strategy for students having difficulty: "Each design is made from familiar geometric shapes. Find these shapes and try to figure out how they are put together."

As students describe their figures, you can introduce correct terms for them. As you use them naturally as part of the discussion, students will begin to use and recognize them.

Quick Image Dot Patterns Use the Quick Image Dot Patterns transparency. The procedure is the same, except that now students are asked two questions: "Can you draw the dot patterns you see? Can you figure out how many dots you saw?"

When students answer only one question, ask them the other again. You will see different students using different strategies. For instance, some will see a multiplication problem, 6×3, and will not draw the dots unless asked. Others will draw the dots, then figure out how many there are.

Using the Calculator You can integrate the calculator into the Dot Pattern Quick Images. As you draw larger or more complex dot patterns, students may begin to count the groups and the number of groups. They should use a variety of strategies to find the total number of dots, including mental calculation and the calculator.

Related Homework Options

- **Creating Quick Images** Students can make up their own Quick Images to challenge the rest of the class. Talk with students about keeping these reasonable—challenging, but not overwhelming. If they are too complex and difficult, other students will just become frustrated.
- **Family Quick Images** You can also send images home for students to try with their families. Instead of using the overhead projector, they can simply show a picture for a few seconds; cover it up while members of the family try to draw it; then show it again, and so forth. Other members of the family may also be interested in creating images for the student to try.

The following activities will help ensure that this unit is comprehensible to students who are acquiring English as a second languge. The suggested approach is based on *The Natural Approach: Language Acquisition in the Classroom* by Stephen D. Krashen and Tracy D. Terrell (Alemany Press, 1983). The intent is for second-language learners to acquire new vocabulary in an active, meaningful contex.

Note that *acquiring* a word is different from *learning* a word. Depending on their level of proficiency, students may be able to comprehend a word upon hearing it during an investigation, without being able to say it. Other students may be able to use the word orally, but not read or write it. The goal is to help students naturally acquire targeted vocabulary at their present level of proficiency.

We suggest using these activities just before the related investigations. The activities can also be led by English-proficient students.

Investigation 1

forward, backward, right, left, turn, steps, robot

1. Move forward, backward, to the right, and to the left. Identify each move as you do so.

   I go forward. I go backward. I turn to the right. I turn to the left.

2. Have students follow your actions as you call out commands.

   Go 3 steps forward. Turn to the right. Go 2 steps backward. Turn to the left.

3. Show and identify a picture of a robot. Point to the robot and then to the students. Tell them that they are going to be robots who follow your commands. Have the student "robots" follow the same short directions you gave earlier, but this time without following your actions.

highest, lowest, in between

1. Write the following numbers on the board:

   2, 8, 12

   Point to the 2 and identify it has having the *lowest* value of these three numbers; identify the 12 as having the *highest* value; identify 8 as a number with a value *in between* 2 and 12.

2. Write another series of numbers on the board, for example:

   23, 35, 42, 61

   Point to one number in the series and make a true or false statement about its value. If the statement is true, students should raise both hands and nod affirmatively. If it's false, they should cross their arms and shake their heads negatively.

   23 is the *lowest* number in this series. (T)

   23 is an *in-between* number in this series. (F)

   61 is the *lowest* number in this series. (F)

   61 is the *highest* number in this series. (T)

3. Repeat with other number series.

small, middle-sized, big, giant, baby

1. Prepare and show cutouts of feet of several different sizes. Identify each:

   This is a large foot. This is a small foot. This is a middle-sized foot.

   Ask students to arrange the cutouts in descending order.

2. Sketch on the board (stick figures) and identify a *baby* and a *giant*. Show students a very small and an extremely large cutout of a foot, explaining that the small foot is the baby's and the large one is the giant's. Ask students to place these in their proper place among the other ordered foot cutouts.

3. Challenge students to demonstrate comprehension of these words by following action commands or by responding to questions that require only one-word responses.

   Show me the *giant* foot.

   Is a *giant* foot bigger or smaller than a middle-sized foot?

straight path, route, distance

1. Make a quick sketch on the board:

   Point to the two paths explaining that one is a *straight path* and the other is *not* straight. Identify them as two different *routes* to the same place. Ask students which route is the shortest distance.

2. Challenge students to demonstrate comprehension of these words by following action commands or answering questions that require only one-word responses.

   Walk in a straight path from your desk to me.

   Is the distance from your desk to the door long or short?

   Show me a route from here to the door that does *not* follow a straight path.

Investigation 2

clothes, hats, sleeves, pants

1. Make a quick sketch on the board:

   Identify the *pants*. Ask how many students are wearing pants. Draw a hat on the figure. Ask how many students are wearing hats. Point to the arms, and explain that this shirt has no sleeves. Then add two sleeves to the drawing. Ask students how many are wearing clothing with sleeves.

2. Challenge students to demonstrate comprehension of these words by following action commands.

   Point to your sleeves.

   Point to someone wearing pants.

   Draw a hat.

   Find someone who is wearing the same kind of clothes you are.

Investivation 4

bedroom, kitchen, living room, furniture

1. Use magazine or other pictures to show students a bedroom, kitchen, and living room. Identify each room. Also point out and name typical objects and furniture in each room.

2. Post these pictures. Challenge students to demonstrate comprehension of these words by following action commands or answering questions that require only one-word responses.

   Stand up when I point to the living room.

   Raise your hand when I point to the bedroom.

   Is the couch in the living room or the kitchen?

   Point to a piece of furniture in the bedroom.

Blackline Masters

____________, 19 ____

Dear Family,

Over the next several weeks, our class will be working on measurement in a unit called *From Paces to Feet*. We'll start out measuring distances with "giant steps" and "baby steps." Later on, your child will be pacing off distances in the classroom and at home. After making these informal kinds of measurements, we will start using rulers, yardsticks, and metersticks. We'll be measuring things like people's feet, their paces, and their heights.

Measurement activities include a fun side—such as figuring out how many giant steps it takes to win a game of Mother, May I? Children enjoy playing such games, and they're learning important math at the same time.

The more experience with measurement that children get at school and at home, the better. Encourage your child to estimate and measure distances. Typical questions that might come up at home include these:

How far is it across our kitchen table—and can we really reach that far?

How many children can sit comfortably on our couch? How many adults?

Will that extra bookcase really fit in the kids' bedroom?

These are good questions, and they're also very practical ones! Measurement questions come up a lot in our home lives, and it's exciting for children to be involved with these real-world issues.

Here's how you can help during this unit:

Listen to your child's strategies for measuring.

Involve your child in your own measurement activities—hobbies like sewing or carpentry are a natural for this.

Work together on the measurement activities your child brings home.

Don't worry if your child doesn't use a ruler accurately yet—it's a skill that will develop over time, with more and more opportunities to measure.

Happy measuring!

Sincerely,

Name ______________________________ Date

Giant Steps and Baby Steps

Are you using giant steps or baby steps? ______________________

Count the steps:

From the stove to the front door.

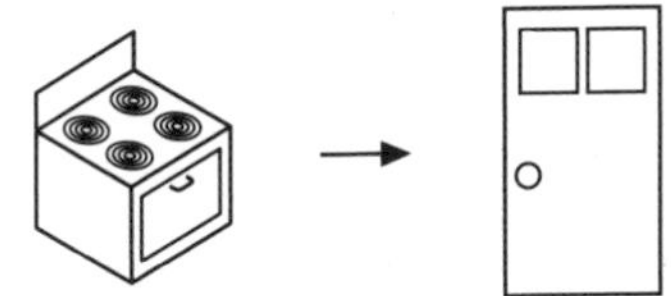

____

From the kitchen sink to the living room couch.

____

From your bed to the bathroom.

____

From the bathroom to the stove.

____

Measure other distances at home. Use giant steps or baby steps. Write the results here.

I used ______________ steps. (Fill in **giant** or **baby**.)

From ________________ to ________________ is ________________.

From ________________ to ________________ is ________________.

From ________________ to ________________ is ________________.

From ________________ to ________________ is ________________.

Name

Date

Student Sheet 2

Pacing and Turning

Write directions for your robot to get to each target.

First target

Location: ____________________

Robot's name: ____________________

____

____

____

____

Second target

Location: ____________________

Robot's name: ____________________

____

____

____

____

Third target

Location: ____________________

Robot's name: ____________________

____

____

____

____

Fourth target

Location: ____________________

Robot's name: ____________________

____

____

____

____

Fifth target

Location: ____________________

Robot's name: ____________________

____

____

____

____

Sixth target

Location: ____________________

Robot's name: ____________________

____

____

____

____

Name Date

Student Sheet 3

Pacing and Writing Directions at Home

1. Write robot directions for each of these routes.

   going from the kitchen sink to your bed

   going from the stove to the front door

   going from the living room couch to the refrigerator

2. Write sets of robot directions for routes that you choose.

   from the ____________ to the ____________

   from the ____________ to the ____________

Name Date

Student Sheet 4

The King's Foot

The carpenter made a new stall for the king, but it was too small to fit the new horse for the princess. Write a letter to the carpenter. Your letter should answer these questions:

Why did the stall end up too small for the new horse?

What could the carpenter do to correct her mistake?

Also make a diagram or picture to show why the stall was too small.

Diagram or picture:

Name Date

Measuring Center Data Sheet

1. How far can you jump?

2. How far can you blow a pattern block?

3. Make a list of the “body benchmarks” you found.

1 inch

6 inches

1 foot

Name

Date

Student Sheet 6

Measure and Compare (page 1 of 2)

1. Measure each of these things. Write down your measurements.
2. Write a sentence saying how much bigger one thing is than the other. Show or tell how you figured it out.

scissors

marker

ruler

pencil

length of table

length of bookshelf

Measure and Compare (page 2 of 2)

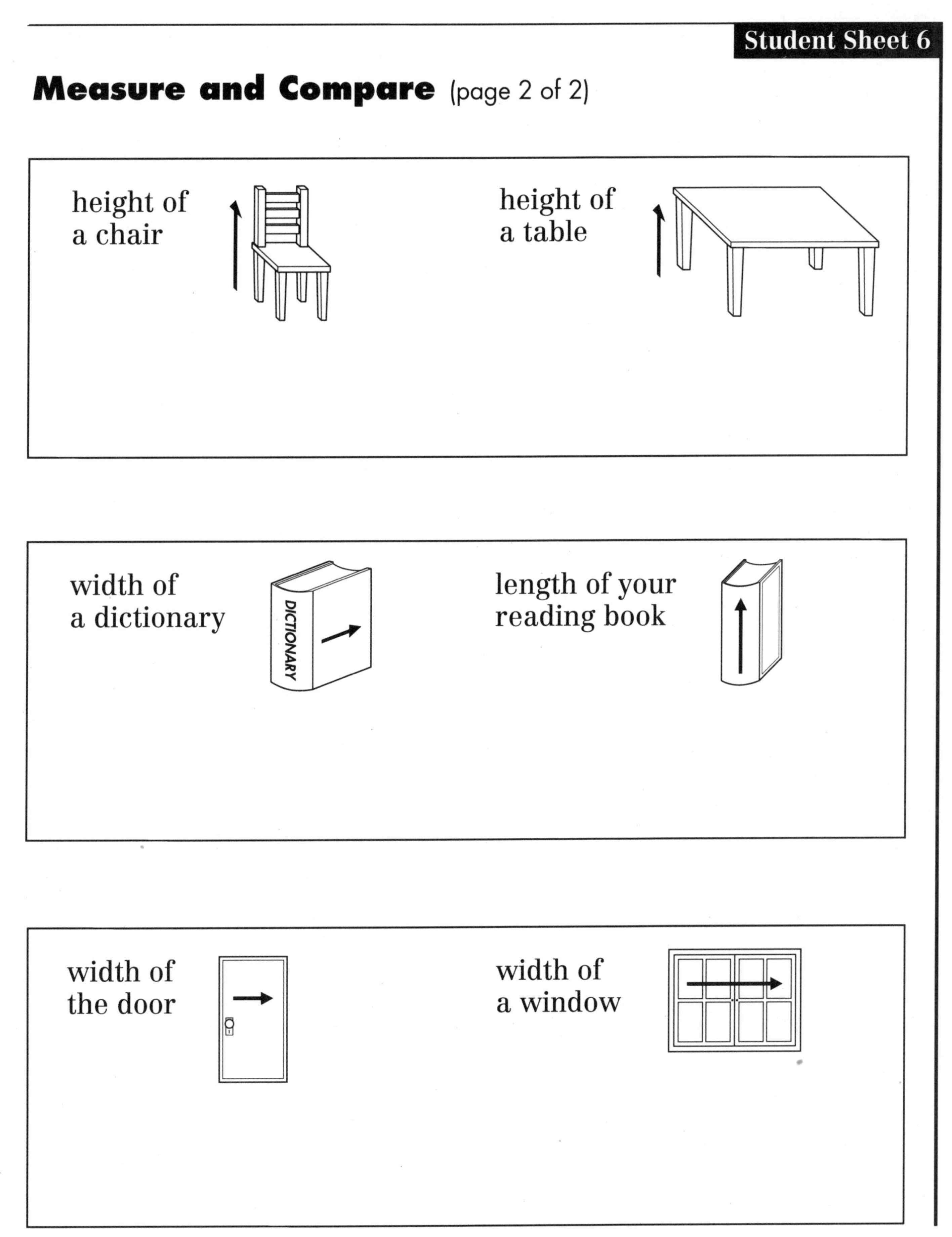

Name

Date

Student Sheet 7

Metric Scavenger Hunt

Things that are 1 meter long and 1 centimenter long

Did you do this scavenger hunt at home or at school? ______________________

Things I found that are about 1 meter long:

Things I found that are about 1 centimeter long:

Name Date

Student Sheet 8

My Sizes in Metric

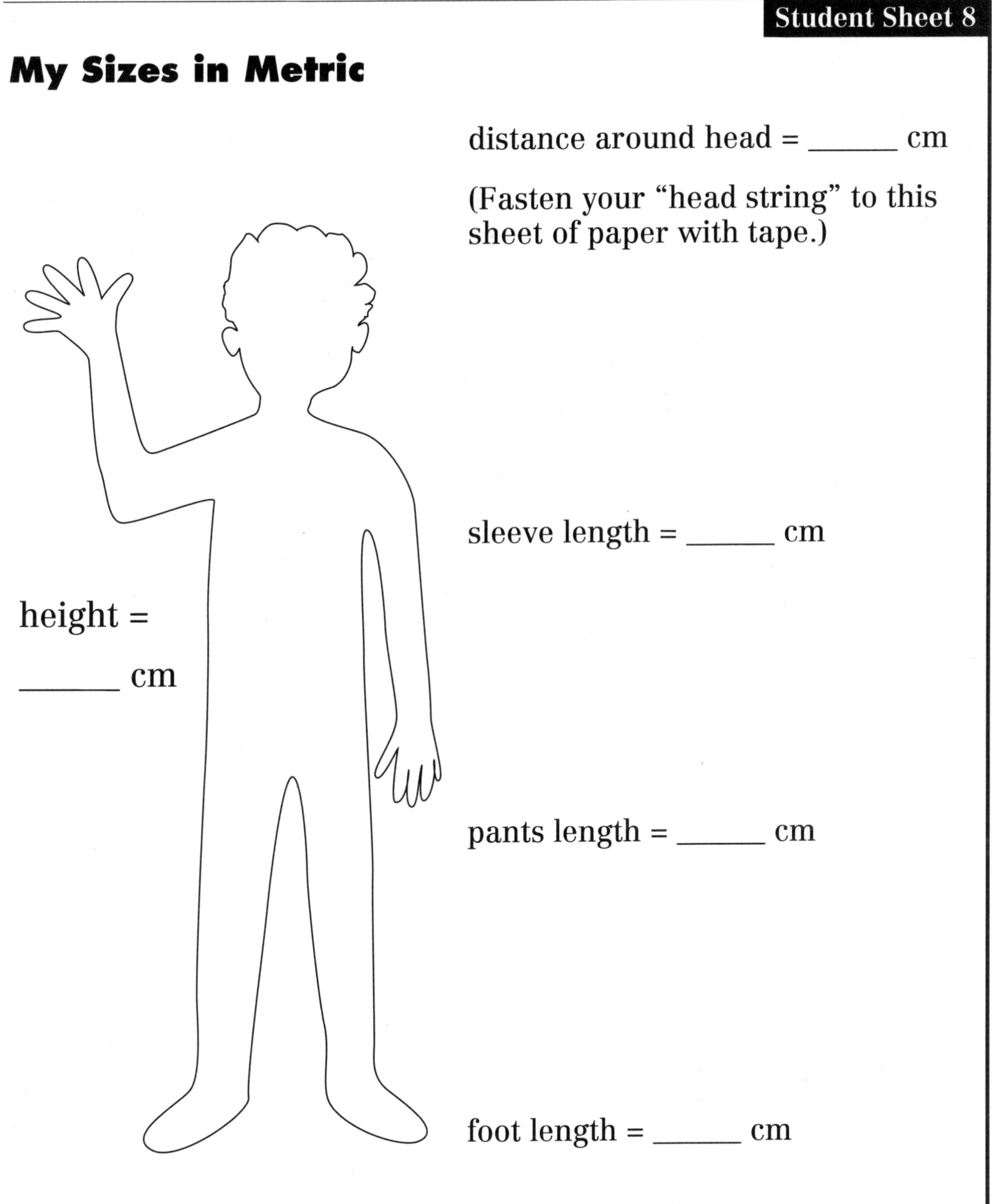

distance around head = ______ cm

(Fasten your "head string" to this sheet of paper with tape.)

sleeve length = ______ cm

height =

______ cm

pants length = ______ cm

foot length = ______ cm

Name Date

Student Sheet 9

Balobby Plans

You are working for a Balobby architect.

For each room that you design, complete one of these sheets.

1. Size of your Balobby: ______________________________
2. What room or place are you making? ____________________
3. What happens in this room? __________________________

____

____

____

____

4. How big is this room? ______ × ______ centimeters
5. What things go in this room and how big are they?

Thing	How big?

Balobby Space Cards

Balobby Bedroom

Design a bedroom for a Balobby child who loves animals. The bedroom is 20 cm × 18 cm. It is filled with different pets. Make tanks and cages that you think are about the right size for the pets.

Remember, the bedroom also needs a place for the child to sleep. There should be something to keep clothes in. Make any furniture that you need.

Make all furniture the right size for a Balobby child.

Futur-kitch

The futur-kitch is a very important place for Balobby families. This is where they have their meals. It's also where they keep their robot. The robot cooks their meals and cleans up.

A futur-kitch is 18 cm × 30 cm. It needs the usual kitchen things—whatever you think should go in a kitchen. Make these things the right size for the Balobbies.

Finally, make the robot. It's 2 cm × 2 cm. It can fit under a counter.

The Grandlobby

Living rooms in Balobbyland are called grandlobbies. A grandlobby is 35 cm × 15 cm. It always has interesting furniture.

Decide what furniture you'll make and what it's used for. Every Balobby family has a big TV, about 8 cm × 2 cm. One other thing—there's a rolling cart filled with snacks (for when they watch TV).

Make plans for all these things. Make other interesting furniture for the Grandlobby, too.

The Teddy Kennel

Balobbies have large teddy bears for pets. They keep them outdoors in a kennel. A kennel is 32 cm × 8 cm.

The kennel should include a house big enough for two teddy bears—they always like to have company.

Decide what else you think the teddy bears need in their kennel. Make plans for these things.

Lobbyschool

At school, the Balobbies usually have 12 students in a class. Their classrooms are 40 cm × 18 cm. Each classroom needs a big rug, where students sit to have meetings. Make the rug big enough for all 12 students to sit on.

What else would the Balobbies need in their classroom, in order to do their work? Make plans. Think about what amount of space and what furniture the Balobby teacher needs, too.

School Fun Room

At school, the Balobbies have an indoor fun room where they can play at recess or after school.

The fun room is 24 cm × 38 cm. It has a small trampoline. It also has things of different sizes to climb over. You may want to make a place to keep snacks. Make these things a good size for the Balobbies.

Balobby Adventure Park

The Balobbies have an outdoor adventure park.
It is 23 cm × 35 cm.

In the park, there are two Balobby tree houses. One is 6 cm × 9 cm. The other is 5 cm × 10 cm. You'll need to make some footbridges to connect these.

There's also a Play Pit. It's big enough for five Balobbies to roll and bounce around with colored balls. Make the Play Pit a comfortable size.

Make other things the Balobbies would like in their adventure park.

Inchstick Pattern

An inchstick is a ruler, 12 inches long, with inch lengths marked off in alternating dark and light squares. Because there are no numbers on an inchstick, students must actually count the units as they make measurements of length, rather than reading off a number that they may not understand.

Durable plastic inchsticks can be purchased from Dale Seymour Publications. You can make your own out of tagboard and laminate them for extra durability. Inch-square graph paper can be used to save time in measuring and marking the inch lengths.

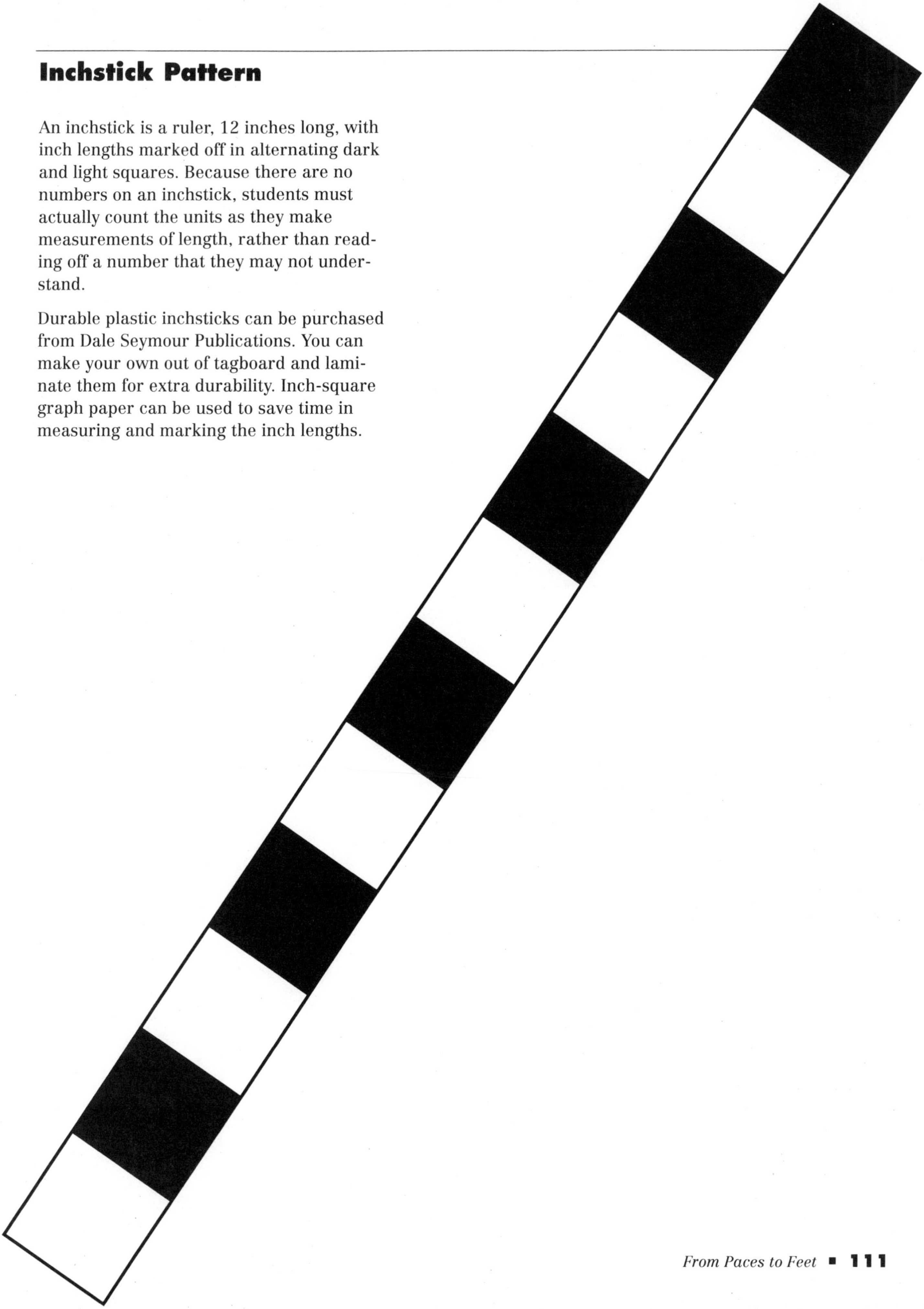

ONE-CENTIMETER GRAPH PAPER

Quick Image Dot Patterns

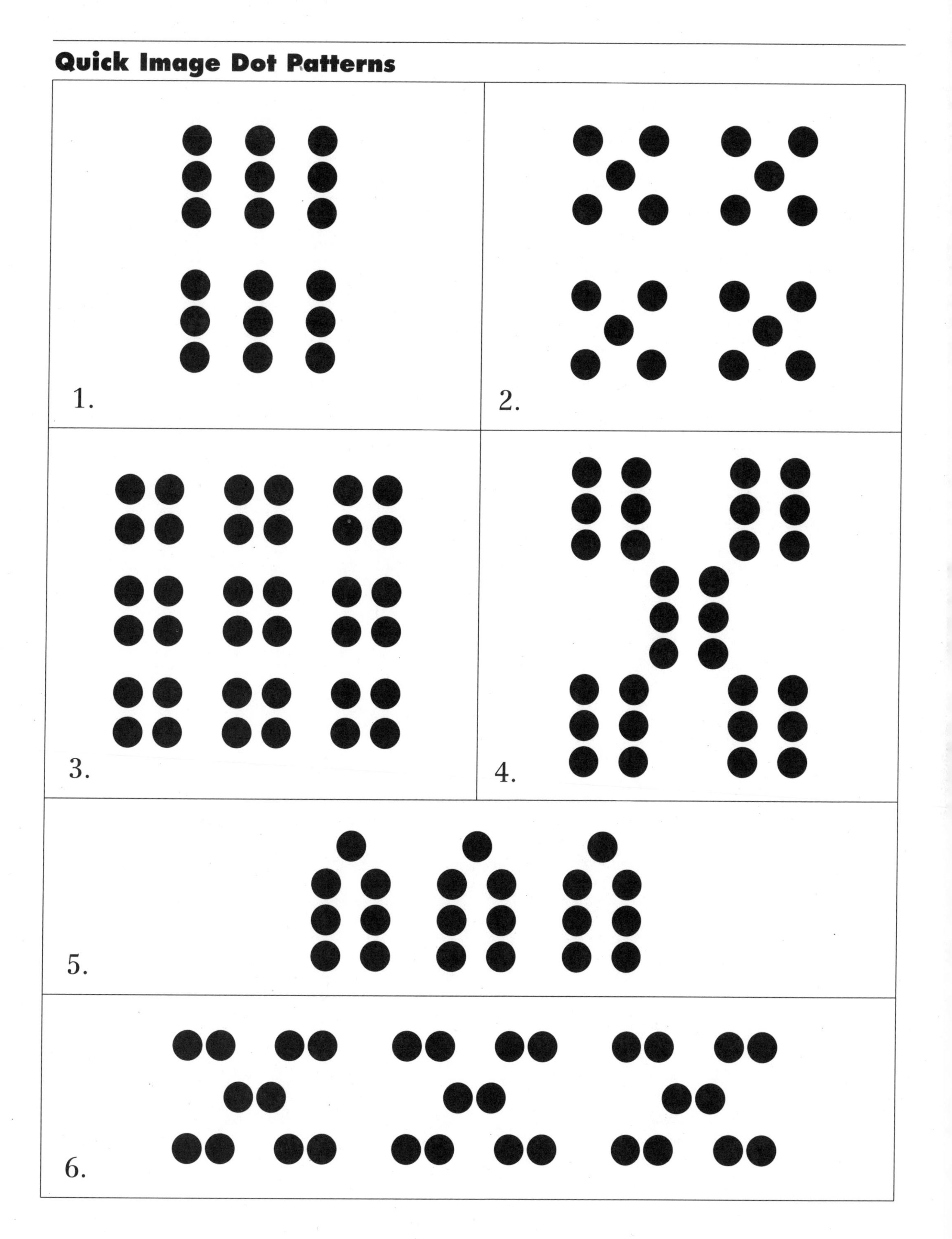

Quick Image Geometric Designs

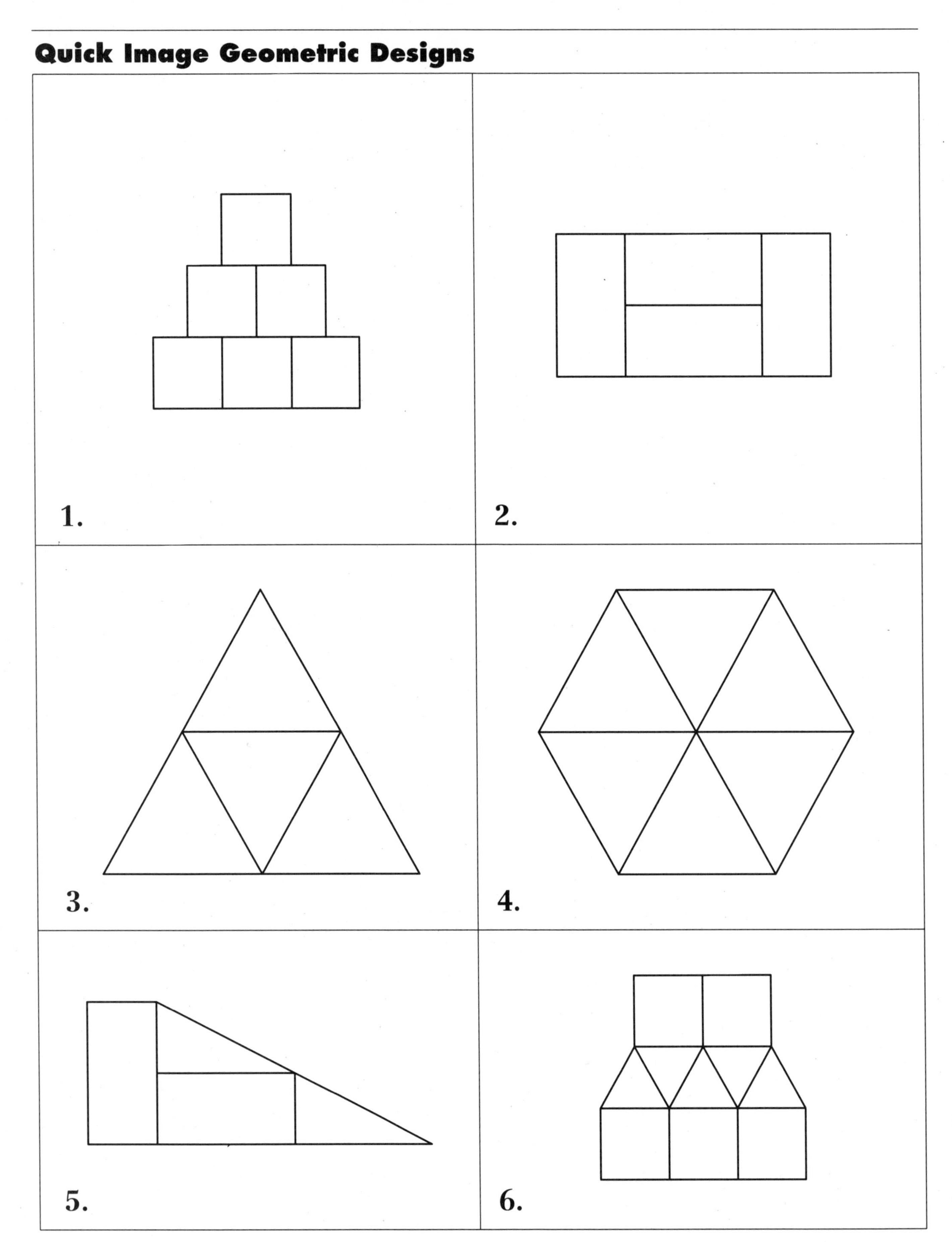